TRANSIT COOPERATIVE RESEARCH PROGRAM

Report 50

A Handbook of Proven Marketing Strategies for Public Transit

TEXAS TRANSPORTATION INSTITUTE
College Station, TX

SOUTH WEST TRANSIT ASSOCIATION
San Antonio, TX

and

UNIVERSITY OF WISCONSIN–MILWAUKEE
Milwaukee, WI

Subject Areas

Public Transit

Research Sponsored by the Federal Transit Administration in Cooperation with the Transit Development Corporation

TRANSPORTATION RESEARCH BOARD
NATIONAL RESEARCH COUNCIL

NATIONAL ACADEMY PRESS
Washington, D.C.   1999

TRANSIT COOPERATIVE RESEARCH PROGRAM

The nation's growth and the need to meet mobility, environmental, and energy objectives place demands on public transit systems. Current systems, some of which are old and in need of upgrading, must expand service area, increase service frequency, and improve efficiency to serve these demands. Research is necessary to solve operating problems, to adapt appropriate new technologies from other industries, and to introduce innovations into the transit industry. The Transit Cooperative Research Program (TCRP) serves as one of the principal means by which the transit industry can develop innovative near-term solutions to meet demands placed on it.

The need for TCRP was originally identified in *TRB Special Report 213—Research for Public Transit: New Directions*, published in 1987 and based on a study sponsored by the Urban Mass Transportation Administration—now the Federal Transit Administration (FTA). A report by the American Public Transit Association (APTA), *Transportation 2000*, also recognized the need for local, problem-solving research. TCRP, modeled after the longstanding and successful National Cooperative Highway Research Program, undertakes research and other technical activities in response to the needs of transit service providers. The scope of TCRP includes a variety of transit research fields including planning, service configuration, equipment, facilities, operations, human resources, maintenance, policy, and administrative practices.

TCRP was established under FTA sponsorship in July 1992. Proposed by the U.S. Department of Transportation, TCRP was authorized as part of the Intermodal Surface Transportation Efficiency Act of 1991 (ISTEA). On May 13, 1992, a memorandum agreement outlining TCRP operating procedures was executed by the three cooperating organizations: FTA, the National Academy of Sciences, acting through the Transportation Research Board (TRB); and the Transit Development Corporation, Inc. (TDC), a nonprofit educational and research organization established by APTA. TDC is responsible for forming the independent governing board, designated as the TCRP Oversight and Project Selection (TOPS) Committee.

Research problem statements for TCRP are solicited periodically but may be submitted to TRB by anyone at any time. It is the responsibility of the TOPS Committee to formulate the research program by identifying the highest priority projects. As part of the evaluation, the TOPS Committee defines funding levels and expected products.

Once selected, each project is assigned to an expert panel, appointed by the Transportation Research Board. The panels prepare project statements (requests for proposals), select contractors, and provide technical guidance and counsel throughout the life of the project. The process for developing research problem statements and selecting research agencies has been used by TRB in managing cooperative research programs since 1962. As in other TRB activities, TCRP project panels serve voluntarily without compensation.

Because research cannot have the desired impact if products fail to reach the intended audience, special emphasis is placed on disseminating TCRP results to the intended end users of the research: transit agencies, service providers, and suppliers. TRB provides a series of research reports, syntheses of transit practice, and other supporting material developed by TCRP research. APTA will arrange for workshops, training aids, field visits, and other activities to ensure that results are implemented by urban and rural transit industry practitioners.

The TCRP provides a forum where transit agencies can cooperatively address common operational problems. The TCRP results support and complement other ongoing transit research and training programs.

TCRP REPORT 50

Project B-13   FY'96
ISSN 1073-4872
ISBN 0-309-06602-6
Library of Congress Catalog Card No. 99-71033

© 1999 Transportation Research Board

Price $44.00

Published reports of the

TRANSIT COOPERATIVE RESEARCH PROGRAM

are available from:

Transportation Research Board
National Research Council
2101 Constitution Avenue, N.W.
Washington, D.C. 20418

and can be ordered through the Internet at
http://www.nas.edu/trb/index.html

Printed in the United States of America

FOREWORD

By Staff
Transportation Research
Board

This report identifies, describes, and assesses proven low-cost and cost-effective marketing techniques and strategies appropriate for use in the transit industry. This is a "how-to" handbook for selecting and implementing such techniques at transit agencies. The target audience is transit marketing professionals, public transit managers, and executives who have responsibilities for marketing transit systems.

Marketing plays a critical role in assisting transit agencies in attracting new riders, retaining existing ones, and ensuring support from the community at-large. To maximize its effectiveness, marketing must be viewed as a comprehensive process through which transit agencies develop and provide transit service and communicate the benefits to their employees, patrons, and the general public. Marketing techniques that are both low-cost and cost-effective are needed by transit agencies and may be crucial to their viability. Transit agencies currently use a variety of low-cost techniques. Thus, there is a need to identify, assess, and share the proven strategies so that they may be adopted throughout the transit industry.

Texas Transportation Institute, in association with the South West Transit Association and the Center for Transportation Education and Development at the University of Wisconsin–Milwaukee, prepared the handbook for TCRP Project B-13. To achieve the project objective of identifying proven marketing strategies to implement at transit agencies, the researchers identified and described low-cost and cost-effective marketing techniques currently used at large, medium, and small, urban and rural transit agencies throughout the transit industry. The complete range of low-cost marketing activities includes traditional, broad marketing activities such as pricing, promotions, advertising, planning, and service delivery targeted at specific submarkets. Further, a method was developed to define the criteria that would be used to assess and select creative and promising marketing techniques. Selection of promising practices was made on the basis of cost, cost-effectiveness, ease of implementation, community support, and staff time required to implement the marketing program. A general overview of each strategy is provided. The overview includes a basic description of the strategies, the objective of its implementation, the resources necessary, the time required, the results of the project, any suggested adaptation or refinements, and when the project was implemented. The size of the transit agency implementing the project is indicated by fleet size.

Also provided in this handbook are summary materials on general principles of marketing public transit. Included are checklists and forms to make it easier for the public transit manager to incorporate solid principles of marketing and public relations.

CONTENTS

AUTHOR ACKNOWLEDGMENTS

The research team wishes to acknowledge the many contributions of the public transit marketing professionals and other transit system executives throughout the United States and Canada who submitted candidate marketing strategies and projects for the study effort. The suggestions and comments of the TCRP B-13 Panel were also invaluable to the research team in refining the "how-to" document.

This work was sponsored by the Federal Transit Administration and conducted in the Transit Cooperative Research Program, which is administered by the Transportation Research Board of the National Research Council.

The research performed and the development of this "how-to" document were performed under TCRP Project B-13 by the Texas Transportation Institute (TTI)/Texas A&M Research Foundation (TAMRF), South West Transit Association (SWTA), and the Center for Transportation Education and Development (CTED) at the University of Wisconsin-Milwaukee (UWM). The Texas A&M Research Foundation served as the contractor for the study while the work undertaken by SWTA and UWM was performed under a subcontract with TAMRF.

TTI's Ms. Cinde Weatherby, an associate research scientist, was the principal investigator and performed general supervision of the project. Major assistance in the drafting of the "how-to" document was performed by Mr. Todd Carlson, a TTI research associate. Other key staff working on the project were SWTA Executive Director Carol Ketcherside and Mr. David Cyra, former executive director of the UWM's CTED and principal of Cyra Engineering Transportation Training Consulting. Graphics design development and document formatting was performed by Ms. Debbie Murillo, TTI commercial artist. Additional support was provided by other TTI staff members; Ms. Kelly West, associate research editor, Mr. Mark Anthony Posada, research associate, Ms. Michelle Walker, assistant research editor, and Mr. Bernie Fette, assistant head, Information and Technology Exchange Center.

INTRODUCTION

Introduction

Transit agencies of all sizes face increased budget pressures and the need to do more with less. To maintain service levels, systems must also continue to attract new riders and retain existing ones, as well as ensure support from the community at large. The use of marketing principles and strategies is a significant tool for transit agencies in meeting these goals. Indeed, to maximize the impact of marketing, it must be viewed as a comprehensive process that is well planned, monitored, and evaluated. In a greater sense, the use of marketing techniques and strategies incorporate an emphasis on customer service throughout the transit organization and its relationships and communications with all of its "publics."

As our society becomes more and more a service-based economy, public transit entities (as well as most other public entities) have realized the importance of public image and quality communications. The transit industry has realized that it is not unlike any other industry or business sector when it comes to customer relations. Some of the most successful public transit systems have adopted an approach to marketing of services that does not differ from any privately owned or operated service. As the director of marketing of one of the more successful public transit systems said at a recent presentation to public transit marketing professionals on the system's marketing efforts, "stand up right now and shake off all of that public sector attitude — you are no different than any private sector service — be creative and positive and sell yourselves."

Transit systems of all sizes can benefit from paying attention to marketing theory and principles. Marketing theory says that there are a set of controllable variables that can be used to influence responses by buyers — product, place, price, and promotion. Good marketing, in the private and public sector, calls first for attention to design of a good product or service to fit consumer needs. Other tasks include determining the appropriate distribution channels (place) and price (fare), before supporting those tasks with a promotional program.

Applying these principles to public transit, marketing should be considered as service is developed and planned (routes, schedules, etc.), as it is distributed (fare media, sales outlets) and priced (fares, discounts, etc.), as well as in the traditional sense of how it is promoted (awareness campaigns, advertising, etc.).

In addition to following general marketing theory and principles, a public transit agency must also be cognizant of "people" as an integral component of the marketing mix. Services must adapt to the changing needs of the customers -- for example, the increased need for accessible transport by an aging population. Message strategies also often focus on behavioral consequences - aimed at changing long-established habits. Some of the other challenges faced in public transit marketing include the following:

- explaining subtle or "invisible" benefits (increased transit use equaling less pollution and congestion),
- calling attention to the beneficiaries (the environment, health care costs),
- explaining long-term benefits,
- dealing with the public scrutiny of taxpayers or elected officials,
- dealing with multiple publics (the "public," politicians, administrators, other agencies),
- dealing with limited opportunities for modifying services (routes, schedules), and
- marketing the same services to a variety of groups (for example, senior citizens and students).

This handbook is not meant to provide all of the answers and suggestions needed to totally incorporate marketing principles into your institution. However, we have provided some hints on basics that are appropriate for use in the industry. We have also provided listings of resources for further information on specific topics. The guide was developed to especially assist rural and small urban transit systems that may not have full-time professionals assigned to marketing tasks. The information provided on specific projects can, however, be of interest and benefit to transit marketing professionals or managers of any type or size of system.

Our quick summary of components of a successful marketing program in this chapter includes the following:

- a review of the importance of customer service,
- some suggestions for planning a marketing program,
- some suggestions for evaluating the marketing program,
- an overview of fostering partnerships with the corporate world,
- some suggestions for handling media relations, and
- a checklist for planning an event.

Fostering Consistent Quality Customer Service

Customer service is a topic that has received a good deal of attention in the popular media. Visit any large bookstore to find books in the business "self-help" section on improving customer service. There are a number of books in that genre that are written by Ron Zemke and Kristin Anderson and incorporate "knock your socks off service" in the titles. That series of books has been found especially useful to a number of transit managers around the country. In *Sustaining Knock Your Socks Off Service* and *Managing Knock Your Socks Off Service*, the authors note that superior service is created through a combination of eight tactics and practices, as follows:

1. Finding and retaining quality people
2. Knowing their customers intimately
3. Focusing their units on organizational purpose
4. Creating easy-to-do-business-with delivery systems
5. Training and supporting employees
6. Involving and empowering employees
7. Recognizing and rewarding good performance and celebrating success
8. Setting the tone and leading the way through personal example

The authors also point out possible major barriers to achieving high-quality customer service, as follows:

1. Inadequate communications between departments
2. Employees not rewarded for quality service or quality effort
3. Under staffing
4. Inadequate computer systems
5. Lack of support from other departments
6. Inadequate training in people skills
7. Low morale; no team spirit
8. Bad organizational policies and procedures

As you read through these lists of tactics and principles and possible barriers, think about your own transit organization. The common thread in both lists is the employee. Some public transit systems are beginning to tackle this issue head-on by changing the way they hire vehicle operators. The new emphasis is on personality and people skills, rather than technical driving ability. Assuming that driving skills can be taught, but developing friendliness and people skills are a bit more difficult to achieve.

The major point is that customer service and providing perceived value to the consumer should be woven throughout the entire structure of an organization - not just emphasized in a marketing campaign. **Customer service is an attitude.** Marketing research can assist in assessing it, but the marketing plan is just one place in a system's management structure that should continually consider customer service. Because of the importance of each individual employee to the organization's success, internal communications and effective organizational principles are also imperative. Some examples of internal programs are provided in the Internal Promotions Category.

General Marketing Principles

Developing a Marketing Plan — Planning to be Effective

Planning is the foundation of a successful marketing program. Transit systems don't put service on the street without a plan (a schedule, routes, service standards) or a budget, and the requirements for a marketing program are no different. Marketing planning will allow you to establish your goals, develop a course of action, and describe a methodology for evaluating the program's success and providing information to be used in the next planning cycle.

The Colorado Association of Transit Agencies, with support from the Colorado Department of Transportation and the Rural Transit Assistance Program, sponsored development of The Marketing Cookbook — Recipes for Success. The document was completed by the firm Communique, LLP, of Aspen, CO. Copies of the document may be acquired, at no charge, by sending a request and a self-addressed mailing label to the Colorado Association of Transit Agencies, 225 East 16th Avenue, Suite 1070, Denver, CO 80203.

The Colorado guidebook provides a very good introduction and summary of using strategic research, dealing with the media, advertising, public relations, crisis communications, evaluation, and planning. Included, for example, are detailed descriptions of the types of advertising to consider — newspaper, magazines, yellow pages, brochure distribution, direct mail, outdoor advertising, point of sale, as well as types of broadcast media - television, cable, radio, and Internet.

With the sponsor's permission, we have included a series of worksheets in the pages that follow from the "cookbook" that will walk through the planning process. Completion of the worksheets will establish the framework of an effective program. Scanning the projects and strategies included in this how-to book can be helpful in developing the action plan (Worksheet #5) for your marketing program.

Worksheet #1: Situation Analysis

This is a look at your system and community. The goal is to identify both facts — the riders, non-riders, past successes, upcoming events, community demographics, etc. — as well as perceptions — what people think of us and our service. Much of this information will be available in your current transportation development plan (TDP) or other market research.

You will want to identify information about the system and the service area. You may well need additional sheets.

For the system, you should need:

What were ridership trends, by segment and in total for the past year or two?

Which routes/services are popular, and why?

Which routes and services need ridership, and which are capacity-constrained?

How do customers now learn about the system? What sort of materials are available, and how are they distributed?

Who now rides the bus?

Who doesn't ride the bus, and why?

Are there service plans which should be considered?

What sort of service amenities are there (shelters, telephones, etc.)?

For your service area, you should record:

What is the population in the service area?

What were growth trends in the last year or two?

What are growth projections?

Where are the population densities high?

Where are they low?

What are the trends in traffic?

What are the community's goals?

What role does the transit system play?

What is the political environment?

How do your customers perceive you?

How do elected officials and other influentials perceive you?

Worksheet #2: Problems and Opportunities

Based on the information gathered about the system and the service area, what are the main obstacles and opportunities facing your system? This should list all possibilities, which may be discarded or refined into goals on Worksheet #3.

Problems/Obstacles to Success:

Opportunities:

Worksheet #3: Goals and Objectives

Simply put, these are statements of what you want to accomplish within a specific time frame. Goals must be specific and measurable. Similarly, they should be attainable. What is realistic and appropriate should be determined by the situation analysis.

In transit, goals can usually be expressed in terms of ridership, revenue, or image. In fact, you may have goals for each of these categories, such as:
• Increase ridership on route x by 3% by year-end, compared to the previous year-end.
• Increase transit pass sales by 6% by year-end, compared to the previous year-end.
• Increase awareness of the transit information line by 8% by a month and year as compared to the last time that data was available.

Goal #1

Goal #2

Goal #3

Goal #4

Goal #5

Worksheet #4: Strategies

If a goal is what you want to achieve, a strategy is how you will achieve it. For example, if you set a goal of increasing awareness of your information line, strategies may include print, radio and/or television advertising, placing the phone number on all fleet vehicles, and printed inserts placed into bank statements.

Generally speaking, you should develop several strategies to help reach each of your goals. While some strategies will be specific to one goal area, other strategies will overlap goal areas, so should be listed in each appropriate strategy worksheet.

[A separate sheet should be completed for each goal.]

Goal #1

Strategy A:

Strategy B:

Strategy C:

Strategy D:

Worksheet #5: Action Plans

As with any planning process, the plan is only as good as its execution. Action Plans are the specific activities you will undertake to fulfill your strategies. It is recommended that projects be outlined and scheduled onto a calendar for the year. This provides the simplest framework for managing your marketing program over the course of the year.

A useful tool in developing your action plans is the removable adhesive notepad. Write each step involved in achieving your goals and strategies on a different note, and place it on a board. You can then add and delete steps, and move them around into the final order, helping to establish a time line and ensure that nothing slips through the cracks. These can then be entered onto a plan calendar that will remind you of upcoming deadlines throughout the year.

Work Project:

Purpose:

Description/time lines:

Additional Resource

The Colorado marketing "cookbook" was written specifically with public transit systems in mind; however, there are others that have been successfully used by non-profit organizations. One example of such a workbook, that has been incorporated into a widely used transit marketing course is the Marketing Workbook for Nonprofit Organizations by the Amherst H. Wilder Foundation. Written by Gary J. Stern, the workbook also provides an array of worksheets for planning and executing a marketing program. Copies of the workbook may be acquired by contacting the Foundation at 800/274-6024, or by writing to Amherst H. Wilder Foundation, Publishing Center, 919 Lafond Avenue, St. Paul, MN 55104. Copies of the workbook are $25.00. Permission to reproduce the worksheets is granted in the document. Volume discounts are also offered.

Keeping Copyright Law in Mind

While this manual encourages the liberal sharing of ideas between public transit agencies and systems, copyright laws should also be kept in mind when borrowing from others - especially from any private entity.

The principle of "Fair Use" allows you to take facts freely; however, the expression belongs to the author. By judicious paraphrasing, authors are using facts without using the original author's expression.

While copyright law does not define the exact limits of fair use, here are some suggested questions that you might consider before publishing copyrighted material:

1. Is it going to be used for profit?
2. What is the nature of the work from which quotes are taken?
3. Could it cause economic damage?
4. Relative to the total material, what percentage would be used?

Some suggested limits to use, without asking permission, is 300 words if it is from a book-length work; two lines from poetry; or 10 percent of a letter.

Under the U.S. Copyright Law of 1978, authors possess certain rights automatically upon creation of materials, even if there is no registration with the Copyright Office, although that is recommended. Original work should show a credit line and protect it by displaying the (c) symbol, with a name and year that the work was completed.

These laws apply to all manner of materials - from training materials to published works. In general, documents created for public transit systems are not to be used for profit. If there is any doubt, the author can be contacted to grant permission for using copyrighted materials.

Additional information on copyright can be obtained by contacting the Copyright Office Hotline at 202/287-9100 or writing to Copyright Office LM455, Library of Congress, Washington, D.C. 20559 and requesting Circular 2, Publications on Copyright. A wealth of information is available on the subject by accessing the U.S. Copyright Office in the Library of Congress through the Internet. The home page address is http://lcweb.loc.gov/copyright/.

In Canada, copyright is dealt with in a very similar manner. The Canadian Copyright Act provides that any "fair dealing" with a work for purposes of private study or research, or for criticism, review or newspaper summary is not infringement. However, in the case of criticism, review, or newspaper summary, the user is required to give the source and author's name, if known. Copyright issues are handled by the Copyright Office, a part of a larger agency called the Canadian Intellectual Property Office (CIPO), which comes under Industry Canada. Additional information on copyright can be obtained by contacting that office through the CIPO at 819/997-1936, or writing to Copyright Office, Canadian Intellectual Property Office, Industry Canada, 50 Victoria Street, Place du Portage, Phase I, Hull, Quebec, Canada, K1A 0C9. The Internet address is: http://cipo.gc.ca.

Evaluating the Marketing Program

While the emphasis on marketing at public transit systems has appeared to have heightened in the past decade, the existence of an evaluation program for the marketing program is still very rare. Many systems have very limited resources available to implement the marketing strategies, much less measure the impact of them. However, there are some fairly simple measurements that can be noted to assist in evaluating the marketing program.

As mentioned in the section on developing the marketing plan, the plan should be written with evaluation in mind. Clearly stated goals and objectives should be as specific as possible, and measurable. The primary indicators to be used are ridership, revenue, and image.

Finding Other Resources Mentioned in This How-To Book

Most of the documents mentioned in this "how-to" book are available through the U.S. Department of Transportation's Technology Sharing Program (TSP). These documents were developed with direct or indirect support of federal funds. Single copies of in-stock TSP reports are available at no cost through the main on-line catalog accessed through the Internet at http://www.tsp.dot.gov/. All TSP reports are archived through the National Technical Information Service (NTIS) in Springfield, VA. When reports are no longer available through DOT sources, reports may be purchased through NTIS. Archived TSP reports may be browsed through the TSP web page, and a direct link to NTIS is provided. NTIS may be contacted directly at NTIS Sales Desk, Monday through Friday, 8:30 a.m. to 5:00 p.m. Eastern Time, at 800/533-6847 or 703/605-6000, fax 703/321-8547. Additional NTIS ordering information is available at the NTIS website - http://www.fedworld.gov/ntis/ordering.htm.

Ridership

The main outcome being measured is ridership on a route or service, compared to ridership for that same route or service during a previous time period. The units are usually periods of a month, quarter, and year. If it is a new route, or service, ridership may be benchmarked against the projections of ridership rather than historical numbers. It is important that the data be as accurate as possible.

Revenue

The measurement of revenue is also fairly simple. Measure the amount of money generated on a given product (such as a certain fare pass) or service (fare box recovery). Benchmarks typically used include:

- projections made in the marketing plan (revenue compared to expectations)
- historical trends (revenue generated this month/quarter/year compared to previous time periods)
- revenue per passenger (total passengers/total revenue; for a route or the total system)
- revenue per hour of service (total hours/total revenue; for a route or the total system)
- revenue per mile of service (total miles/total revenue; for a route or the total system)

Image

Image is more difficult to measure because the data is not already being collected as a matter of daily record. It has to be collected through market research, and there should be a baseline established as a benchmark so that the results of the marketing efforts may be measured against the baseline.

Ongoing Research Programs

An ongoing research program can include both qualitative research and quantitative studies. Examples of qualitative research could include focus groups and one-on-one interviews. Both of these types of research can provide insights that might be missed in a quantitative survey. They can also be very useful in surfacing issues and questions to be used in a quantitative survey.

Examples of quantitative research are:

- on-board surveys
- intercept surveys
- telephone surveys
- mail-back surveys

Below are some suggestions about each type of survey, taken from The Marketing Cookbook - Recipes for Success.

On-Board Surveys

Surveying existing passengers is the easiest, most cost-effective form of quantitative research for transit systems. Surveys can be issued to each passenger as they board, and collected as they disembark. Depending on the number of surveys collected, you may elect to tabulate every second or third survey to save on time and cost of data entry, while still arriving at a statistically sound sample. [A sample on-board survey is provided at the end of this section.]

Following are a few recommendations about on-board surveys gleaned from other transit operators' experience:

- Survey passengers in one direction only (inbound or outbound)
- If possible, offer a passenger free fare to assist the driver with distribution and collection of surveys
- Survey passengers on more than one day of the week (for example, Wednesday and Saturday) to get a more representative sample
- Prior to the survey day(s), use media releases and public service announcements to inform the public about the survey

Intercept Surveys

Intercept surveys are brief interviews (one to five minutes) conducted in places with high traffic volume, such as shopping or pedestrian malls. Intercept surveys are a cost-effective way to reach both users and non-users. However, results from intercept surveys do tend to be skewed slightly toward women and middle-income respondents, since statistically this is the majority of shoppers in America. Collecting surveys on a couple of different days of the week (for example, Wednesday and Saturday) and in several different locations will help provide a more representative sample. If you provide specialized services, senior centers, doctor's offices, and clinics can be useful survey sites.

Following are a few recommendations about intercept surveying:

- Think through where you will conduct the surveys, as the venue may affect respondent demographics.
- If possible, select attractive interviewers, and have them dress appropriately for the venue(s) in which the interviews will be conducted.
- Request interviews of a broad cross-section of respondent groups - men, women, older, younger, etc.
- Prior to the survey date, contact a random group of people living in the area by telephone and ask them to stop by to complete the survey, as this will eliminate some of the shopping mall demographic bias.
- If possible, offer a premium (for example, a free movie ticket or free bus pass) to help motivate respondent participation.

Telephone Surveys

Telephone surveys are, obviously, conducted over the telephone with either users or non-users residing in the transit operator's service area. Although slightly more expensive to conduct than intercept surveys, telephone questionnaires are very effective for measuring awareness and image of a transit system. Respondent lists can be purchased from a variety of sources, including the phone company, and targeted to many geographic, economic and demographic factors.

Following are a few recommendations about telephone surveying.

- If possible, share the cost of a telephone survey with related public service organizations such as the housing authority, the city or county planning office, or others interested in the statistical data that you are seeking such as a chamber of commerce.
- Make sure the questions are clear and concise.
- It is useful to test the survey questions prior to execution. Test the survey on people in your organization as well as others.
- Prior to the survey day(s), use media releases and public service announcements to inform the public.

Mail-Back Surveys

An alternative to telephone surveys is to distribute surveys by mail, with a postage-paid return envelope. As with telephone surveys, respondent lists can be purchased from a variety of sources, including the phone company, based on many geographic, economic and demographic factors. The advantage of mail-back surveys is that more complex questions such as "rank the following..." can be asked. On the other hand, mail surveys can be expensive, because a large number of surveys must be mailed out to ensure adequate response. Typical return rates on direct mail is one to two percent. This calls for a large number of surveys to be issued to get back a statistically significant return.

Following are a few recommendations about direct mail surveying:

- Offer incentives and bonuses for completing the survey, such as random prizes;
- Use postage-paid mail permits for return postage, so that postage is paid only on returned surveys; and
- Prior to the survey day(s), use media releases to inform the public of the survey

Use of Information Gained from Research

The information from the initial survey efforts will be very helpful in establishing priorities in the marketing plan. Once the plan is developed, and strategies implemented (with the type of measurements to judge success in mind in advance), the results of the next survey efforts can demonstrate the impact of the strategies. These results will then be very helpful in developing the next marketing plan. The evaluation program should be a consistent part of the annual marketing plan, providing the research each year for updating and improving upon the marketing plan.

SAMPLE PASSENGER SURVEY

We are conducting a study regarding our transit services, and would appreciate your taking a few minutes during your trip to complete this survey.

1. How often do you ride transit?
   - ☐ Less than once a month
   - ☐ 1 to 4 times a month
   - ☐ 1 to 2 times a week
   - ☐ 3 to 5 times a week
   - ☐ 6 or more times a week

2. How long have you regularly ridden transit?
   - ☐ First time
   - ☐ Less than 1 month
   - ☐ 1 to 3 months
   - ☐ 4 to 6 months
   - ☐ 7 to 12 months
   - ☐ 1 to 3 years
   - ☐ 4 to 6 years
   - ☐ 7 to 10 years
   - ☐ More than 10 years

3. How likely is it that you will continue to ride transit?
   - ☐ Very likely
   - ☐ Somewhat likely
   - ☐ Not very likely

4. What is the purpose of this trip?
   - ☐ Work
   - ☐ Shopping
   - ☐ Medical
   - ☐ School
   - ☐ Personal business
   - ☐ Social/recreation

5. How did you get to this bus/van?
   - ☐ Walking less than 3 blocks
   - ☐ Walking 3 or more blocks
   - ☐ Was driven to the stop
   - ☐ Drove self to stop
   - ☐ Transferred from another route
   - ☐ Rode a bicycle
   - ☐ Other (Please specify):_____

6. Did you have a car available for this trip?
   - ☐ Yes
   - ☐ No

7. Do you have a valid driver's license?
   - ☐ Yes
   - ☐ No

8. If you drove to the stop, where did you park?
   - ☐ Designated park and ride lot
   - ☐ General "on street" parking
   - ☐ Other (Please specify):_____

9. How far do you normally travel to work?
   - ☐ Less than one mile
   - ☐ 1 to 3 miles
   - ☐ 4 to 10 miles
   - ☐ 11 to 20 miles
   - ☐ 21 to 30 miles
   - ☐ 31 to 40 miles
   - ☐ Over 40 miles

10. Why did you choose to ride transit? (Check as many as may apply)
    - ☐ Only alternative
    - ☐ No car available for this trip
    - ☐ Avoid traffic
    - ☐ Save time parking
    - ☐ Save money on the cost of travel
    - ☐ Employer provides transit pass
    - ☐ I care about the environment
    - ☐ Other (Please specify): _____

Following are statements about this transit service. We would like to know your opinions. Please check only one answer per statement. You should check the box that corresponds to how strongly you agree or disagree with the statements.

11. It's reliable.
- ☐ I agree very strongly
- ☐ I agree strongly
- ☐ I agree somewhat
- ☐ I don't really agree

12. It's convenient.
- ☐ I agree very strongly
- ☐ I agree strongly
- ☐ I agree somewhat
- ☐ I don't really agree

13. It's as fast as going by car.
- ☐ I agree very strongly
- ☐ I agree strongly
- ☐ I agree somewhat
- ☐ I don't really agree

14. It's economical.
- ☐ I agree very strongly
- ☐ I agree strongly
- ☐ I agree somewhat
- ☐ I don't really agree

15. It's usually on time.
- ☐ I agree very strongly
- ☐ I agree strongly
- ☐ I agree somewhat
- ☐ I don't really agree

16. The drivers are safe.
- ☐ I agree very strongly
- ☐ I agree strongly
- ☐ I agree somewhat
- ☐ I don't really agree

17. The drivers are friendly.
- ☐ I agree very strongly
- ☐ I agree strongly
- ☐ I agree somewhat
- ☐ I don't really agree

18. Route/schedule information is readily available.
- ☐ I agree very strongly
- ☐ I agree strongly
- ☐ I agree somewhat
- ☐ I don't really agree

19. Information is understandable.
- ☐ I agree very strongly
- ☐ I agree strongly
- ☐ I agree somewhat
- ☐ I don't really agree

20. The vehicles are comfortable.
- ☐ I agree very strongly
- ☐ I agree strongly
- ☐ I agree somewhat
- ☐ I don't really agree

21. Transit services should be publicly subsidized (by taxes).
- ☐ I agree very strongly
- ☐ I agree strongly
- ☐ I agree somewhat
- ☐ I don't really agree

How would the following things or events affect your current transit usage?

22. Earlier morning service
- ☐ Definitely ride more often
- ☐ I might ride more often
- ☐ It would have no effect
- ☐ I might ride less often

23. Later evening service
- ☐ Definitely ride more often
- ☐ I might ride more often
- ☐ It would have no effect
- ☐ I might ride less often

24. More frequent commuter service
- ☐ Definitely ride more often
- ☐ I might ride more often
- ☐ It would have no effect
- ☐ I might ride less often

25. More frequent mid-day service
- ☐ Definitely ride more often
- ☐ I might ride more often
- ☐ It would have no effect
- ☐ I might ride less often

26. Fare increase of 50 cents
- ☐ Definitely ride more often
- ☐ I might ride more often
- ☐ It would have no effect
- ☐ I might ride less often

27. Fare decrease of 50 cents
- ☐ Definitely ride more often
- ☐ I might ride more often
- ☐ It would have no effect
- ☐ I might ride less often

28. You are: ☐ male   ☐ female

29. Your age is:
☐ Under 13   ☐ 35-44   ☐ 13-18   ☐ 45-54   ☐ 19-24   ☐ 55-64   ☐ 25-34   ☐ 65+

30. Your occupation is: _____

31. Your total annual household income is: _____

32. Please provide your ideas for improving this service: _____

_____

_____

_____

THANK YOU FOR YOUR ASSISTANCE!!!

Source: The Marketing Cookbook - "Recipes for Success"

Sample Telephone or Intercept Survey

We are conducting a study regarding transit services, and would like to ask you a few questions.

1. What is your usual mode of transportation?
   - ☐ Walk
   - ☐ Ride private bus/van
   - ☐ Other (Please specify): _____
   - ☐ Ride bicycle
   - ☐ Carpool
   - ☐ Ride public bus/van
   - ☐ Drive alone

2. How many miles do you think you normally travel in a day?
   - ☐ Less than 1 mile
   - ☐ 11 to 20 miles
   - ☐ Over 40 miles
   - ☐ 1 to 3 miles
   - ☐ 21 to 30 miles
   - ☐ 4 to 10 miles
   - ☐ 31 to 40 miles

3. Do you usually have a car available?
   - ☐ Yes
   - ☐ No

4. Do you have a valid drivers license?
   - ☐ Yes
   - ☐ No

5. How often do you ride transit?
   - ☐ Never
   - ☐ 1 to 2 times a week
   - ☐ Less than once a month
   - ☐ 3 to 5 times per week
   - ☐ 1 to 4 times per month
   - ☐ 6 or more times per week

6. How would you rate the overall quality of transit services in your community?
   - ☐ Excellent
   - ☐ Poor
   - ☐ Good
   - ☐ Unacceptable
   - ☐ Satisfactory

7. Why do you choose to ride transit?
   - ☐ My only alternative
   - ☐ I care about the environment
   - ☐ Other (Please specify) _____
   - ☐ To avoid traffic / driving
   - ☐ I never ride transit
   - ☐ To save money / time

8. What keeps you from using transit more?
   (Check all that apply)
   - ☐ House of service
   - ☐ Reliability
   - ☐ Personal comfort
   - ☐ Other (Please specify): _____
   - ☐ Frequency of service
   - ☐ Convenience of departure/arrival
   - ☐ Information about routes/schedules
   - ☐ Cost of service
   - ☐ Convenience of stop location
   - ☐ Personal safety

9. How far from your home is the nearest stop?
   - ☐ 3 blocks or less
   - ☐ More than a mile
   - ☐ 3 to 6 blocks
   - ☐ Unsure
   - ☐ 7 to 12 blocks

10. Should transit services be publicly supported by taxes (like policy, fire, and other community services)?
    - ☐ Yes
    - ☐ No
    - ☐ Unsure

11. Do you have any comments or suggestions to improve transit services in your community?

Thank you for your assistance!!!
Source: The Marketing Cookbook — "Recipes for Success"

Additional Evaluation Tools

Because it appears that so little evaluation of marketing strategies is being captured by systems across the country, we have provided below more discussion on the subject of evaluating pricing-related promotions. There are a variety of pricing-related promotion examples included in this how-to book. The suggestions for evaluating pricing-related promotions come from the Public Transportation Marketing Evaluation Manual - Techniques for Data Collection that was prepared under the federal transit technical assistance program. Individuals interested in acquiring a copy of the document may do so by contacting the Technology Sharing Program (see page 192).

Data Collection Techniques for Pricing Promotions

Overview

The marketing mix includes product, place, price and promotion (for public transportation "product" and "place" can be combined into "service"). The "price" term in the marketing mix is not what we're discussing here, which refers to the normal ongoing and regular price of a service or a product. What we ARE discussing is pricing promotions, which are part of the fourth element of the marketing mix - promotion. Pricing promotions are included in this category, along with advertising, publicity, direct sales, graphics and incentives.

Pricing promotions fall under the broader promotional category - incentives. Pricing promotions include a short-term reduction in price in order to increase sales. They can, in the case of public transportation, include

- free rides,
- discounts on passes,
- fare discounts during off-peak hours,
- distribution of coupons good for free or reduced fare rides, and
- merchant discount programs.

The last item mentioned is a program in which riders receive coupons as they board a public transportation vehicle (or purchase a monthly pass) which is, in turn, worth a discount at participating merchants. Although this is not directly a discount on the transportation fare, customers do gain an economic incentive almost equal to a fare, or indeed sometimes more, thus the reason for categorizing it as a pricing promotion.

It is rather arbitrary as to when (how much time) a short-term manipulation in price must be in effect before a pricing promotion becomes a basic change in pricing structure. To keep issues simple, we will simply call a price change (either in the cost of a fare, or an economic return of some sort) a pricing promotion if it is not a basic change in the pricing structure and is viewed as temporary.

Pricing promotions are one of the easiest promotional strategies for which to gather data. This is so because the explicit goal is to increase sales (rides), so the most logical data collection technique is to measure ridership. In contrast, one must measure attitudes, cognitions, knowledge, intimidation and so forth when evaluating consumer information aids and advertising. There are, however, some other dependent variables that one might want to look at beyond ridership when evaluating pricing promotions. In merchant discount programs, for example, it might be appropriate to assess the merchants' satisfaction with, or willingness to stay in, the discount program, or to measure the number of increased customers a merchant received because of involvement.

Data collection techniques are provided that are appropriate for evaluating:

- free and reduced fare programs (no coupon necessary - simply a reduced or free fare when boarding a vehicle)
- pass discounts
- coupons good for fare reductions
- merchant discount programs
- lotteries (each fare paying passenger has odds on winning a prize)

Techniques

Technique:	**Ridership counts**
Apply to:	Free and reduced fare programs, pass discounts, coupons good for fare reductions, merchant discount programs, lotteries
What it measures:	Ridership
How it works:	Ridership is measured by fare box revenue, manual passenger counts, or by automatic passenger counters. It would be ideal to get counts before, during and after the program. It would even be better to also measure ridership on routes on vehicles without the pricing promotions, so one could obtain good concurrent control data.
Advantages:	Strong proof, if appropriate controls are taken, of the impact of the pricing promotion.
Disadvantages:	Day-to-day vacillations in ridership due to uncontrollable variables such as weather may override the effects of the promotion. It is often difficult to get accurate ridership counts at reasonable costs.

Technique:	**Pass sales**
Apply to:	Pass discounts, merchant discount programs, lotteries
What it measures:	Pass sales
How it works:	Through the pass sales or accounting office, document pass sales during a pricing promotional program. Appropriate controls would be necessary (e.g., pass sales before and after the promotional program). Technique directly assesses impact of pass discount programs and merchant discount programs and lotteries when the benefits are contingent upon pass purchase. Pass sales may also go up when a merchant discount or chance on a lottery are contingent on simply boarding a transit vehicle.
Advantages:	Easy data to collect as pass sales are routinely collected by the sales or accounting departments.
Disadvantages:	Pass sales may not directly correlate with ridership changes. Also, it may be hard to obtain "fine grain" data on pass sales (e.g., sales in the mornings, or sales from individual outlets).

Technique:	**Coupon tracing**
Apply to:	Pass discounts, coupons good for fare reductions
What it measures:	Number and type of coupon turned in
How it works:	Different types of coupons, good for a reduction on either a pass

or cash fare are distributed to potential consumers in a controlled fashion (i.e., a control group gets same mailing, etc., but with no coupons). Coupons of various types (color coded) are deposited in the fare box when used and counted daily. Control group is called by phone to measure their bus riding.

Advantages: Simple, easy to administer way to document a variety of pricing promotions in a controlled fashion.

Disadvantages: Some problems include: consumers' negative reaction to coupons, counterfeiting, and coupons jamming the fareboxes.

Technique: **Merchant data**
Apply to: Merchant discount programs
What it measures: Sales by participating stores, discount coupons turned into participating stores, attitudes of participating merchants, and merchant willingness to stay in the program.
How it works: Coupons for discounts at local stores are given to transit customers as they board a vehicle or purchase a transit pass. If the participating merchants will allow it, changes in sales data and/or an accounting of the coupons turned in to their stores would be excellent data. At the least, the merchants' attitudes (over time) in regards to the program or their willingness to stay in the program would be good data.
Advantages: Retail sales data and number of coupons returned are easily collected by the stores.
Disadvantages: A merchant may not want to divulge his or her store's data. A small number of participants (customers) in such a program may count as a success for the transit agency, but a new customer for a small store may be beneficial from the merchant's perspective.

Technique: **Surveys**
Apply to: Free and reduced fare programs, pass discounts, coupons good for fare reductions, merchant discount programs, lotteries
What it measures: Rider, nonrider and merchant attitudes and knowledge of the pricing promotion, and rider reports or rides taken during the program.
How it works: As in the case for consumer information aids and advertising, much information can be gathered via surveys. Surveys for pricing promotions can be administered on the phone, through intercept interviews on the street, in a mall or on a transit vehicle. They may be guided or self-administered. Questions should range from subjective ("Do you like the merchant discount program?") to objective ("Did you ride the bus yesterday?").
Advantages: Can gather a lot of data, from attitudes and knowledge to reported ridership. Can assess some of the more subjective aspects of a pricing promotion, which may tease out subtle effects of a pricing promotion.
Disadvantages: People's answers to surveys may not be accurate indicators of their ridership, and the technique is more expensive than many of the others.

Technique:	**Focus groups**
Apply to:	Free and reduced fare programs, pass discounts, coupons good for fare reductions, merchant discount programs, lotteries
What it measures:	Attitudes and preferences about pricing promotions presented
How it works:	Before, during or after a pricing promotion, a focus group would be a good way to get a subjective overview of the program. Ten to fifteen citizens (riders or nonriders or a mix) are selected from a population segment and attend a one to two hour session to convey their thoughts on the pricing promotion in question. A "neutral" focus group facilitator keeps the group focused on the subject and stimulates discussion without asserting his or her views. A focus group session before the introduction of a pricing promotion could yield valuable data for the design of the program.
Advantages:	Relatively efficient in time and money; technique garners a wide range of information in regards to the pricing promotion.
Disadvantages:	Subjective data may not be indicative of actual impact of a pricing promotion; a dominant member of the focus group could lead other members and bias output.

Developing Partnerships with the Community

Many of the ideas that are presented in this "how-to" manual involve working cooperatively in partnership with the corporate community or other non-profit or governmental entities. Below is an overview of things to keep in mind when developing these sort of partnerships. These suggestions were based heavily on the public relations course presentations of Ms. Alison Ducharme of AD & Associates in Victoria, British Columbia. Ms. Ducharme is a marketing, sponsorship, and event management consultant who also lectures at the University of Victoria.

Corporate sponsorship of the 1960s and 1970s was seen as primarily philanthropic, with funds coming for corporate "donations" budgets. The business of corporate sponsorship has evolved over the years however. Many corporations viewed the corporate sponsorship opportunity from an emotional perspective rather than as being made for good business purposes. The competition for the donations was less intense than today - with fewer not-for-profit organizations seeking support.

Today corporate sponsorship has become a sophisticated business for organizations both large and small to ensure future sustainability. Corporate sponsorship programs are now integrated into the company's marketing plan and more results oriented. With much more competition for the scarce corporate sponsorship dollars, corporations are in a position of choosing the organizations with which they wish to partner. More often than not, the decision is based on the business appeal of the partnership. As a result, non-profit organizations (and public transit agencies) have had to increase the knowledge of their staff in the areas of developing corporate resources.

Acquiring corporate sponsorship is an involved process that can realize big financial gain. However, the process requires planning, just as any other component of the marketing program. A methodical, strategic plan is required prior to recruiting any sponsors. Once this groundwork is accomplished and the organization is clear about its goals and objectives, the success rate will be much greater.

Prior to targeting potential corporate partners, it is recommended that an agency consider the types of organizations with which they wish to be publicly linked in the mind of the consumer. A worksheet is included for use as you consider potential corporate sponsors. Team meetings within your organization can provide opportunities for gathering critical information to consider this issue, as well as discussing your own image strengths.

Corporate executives normally provide a small window of opportunity for delivery of a message and requested sponsorship, thereby requiring a focused, straightforward and concise presentation. The presentation should be a maximum of 20 minutes in length, with the opportunity for questions and clarification beyond that time. The presentation should focus on meeting the objectives of the corporation. As Ms. Ducharme puts it, "bait the hook to suit the fish, not the fisherman."

Prior to developing the proposal and presentation, the agency should do research on the potential sponsors. Knowing the corporate sponsor's objectives allows the proposal to relate directly to that company's bottom line. The proposal should succinctly include an overview of the project being suggested for sponsorship, the business advantage that the corporation will realize from its participation, and a clear summary of the dollars requested.

Once a corporation or other public entity agrees to be a partner with a public transit agency in a project or program, it is essential to maintain ongoing communications. Honest, regular communications is essential to building a strong, long-term partnership. These communications will allow both parties to discuss how the project or program is meeting the short-term and long-term expectations of each, and allow the discussion of future partnering opportunities. In-person or telephone visits remind the corporate sponsor of the desire of the agency to provide them with value for their participation. It can also lead to more creative, mutually beneficial projects.

Value Model for Sponsorship Acquisition Worksheet

Event Description
Write down a few key words that capture your event - what, where, who, why:

Principle Level
On principle, we will not affiliate with companies or organizations who:
(For example, engage in unhealthy/unsafe practices; endorse products/procedures that endanger the environment; actively discriminate)

Preference Level
We would prefer to affiliate with companies or organizations who:
(For example, are proactive in health/wellness/sport/recreation; are proactive in employment equity, access and inclusiveness; are proactive in environmental awareness; possess a positive employment record; are good corporate citizens; are headquartered in our area)

Using Media Relations as a Marketing Tool

Being public entities, most public transit systems will be the object of media coverage. But solid media relations can also be a mighty tool for marketing of the system. Below are some suggestions for media relations that come from the transit marketing handbook, Promotions Publicity and all that Pizazz - Round Two, published in November 1994 by the Ontario Urban Transit Association's Centre for Transit Improvement.

- Decide upon an official transit spokesperson and at least one designated back-up person. Stick to this!
- Know what is going on in your transit system and in your community, now and in the foreseeable future. Above all, know your market. Become involved in organizational meetings. Maintain strong communication with staff. You want to be able to promote good news and to be prepared to effectively release or contend with bad news. After all, how can you promote a heroic deed if you don't know about it? And, how can you be prepared for a negative incident if you are unaware of it?
- Don't just talk a good game, play it. Nothing will discredit you more than to promise something which you are not prepared to deliver. Be timely about putting plans into action, and keep the public informed of those time lines. Really blow the horn once the plans take effect.
- When the media approaches you for a story or an interview, ask what the direction of the story is and what questions the reporter wants answered. Tell the reporter that you will get back with the information prior to the media deadline, and do so.

Unfortunately, however, reporters often call at the last minute, giving you no time to prepare. Convert them! Tell them that the information they will get is, by necessity, minimal or incomplete, and offer them excellent results when and if they give you adequate notice. Then follow through. It is easier for a reporter to write a column with a stack of good, reliable information than with just a few scraps of hearsay. And remember, if you do not know the answer to a question, say so. Offer to find out and get back to the reporter.

- Most things are not worth hiding from the press. They can find out anything that is public information anyway. When you release the information you will at least have some control over what and how it is released.
- Read everything you can about the media. Libraries, community colleges, universities, various marketing associations and electronic bulletin boards are all great sources of information. Attend lectures and seminars about media relations. Take careful notice of key articles in your local newspapers and on radio and television stations. What are the main elements? How could a spokesperson, or someone quoted in an article, have worded something more positively?
- Get to know the local reporters who are writing transportation-related articles. Build a relationship with them. They can be your best friends (and your best and cheapest form of advertising). Invite them into your transit facility (after you thoroughly prepare staff for the event); explain the ups and downs of running your transit system. Offer to answer any questions they may have which come up after the visit.

Call or have someone on your team personally call them every time an opportunity for a transit story comes up. Nothing you do can put you more on the side of a reporter than calling him/her with a negative story (which is going to come out anyway). They will get the "scoop," and you get their trust and are able to minimize damage on a potentially "lethal" story!

Regardless of how good a relationship you have with local reporters, try to keep an open mind. Remember that a negative story usually gets more reaction than a positive one, and no matter what you do, you are going to see some articles that you won't like!

- Fax corrections to mis-quotes right away. If necessary, you can always buy an explanatory ad to help turn around bad or incorrect coverage. Take the initiative to "mend fences." When a reporter has written a negative article, call and offer your side of the story. Keep the tone positive. The only thing that will be achieved by making enemies in the media is to ensure continued bad press!
- Evaluate your press relations on an ongoing basis. Are you getting enough coverage? Is most of it good? Are certain papers or radio stations ignoring you? You will never have everyone "on your side" all of the time, but by building strong media relations you will certainly improve the odds!

If you do not believe that you are getting enough media coverage, do not be shy about proposing ideas to the media. Especially in small media markets, the newspaper is often very receptive to receiving good black and white photographs with caption details and stories about events, actions taken, new services, or any other "news worthy" activity of the agency.

Keep in mind that "human interest" is a big factor. Pictures of new equipment might be run, but a picture of a new bus that includes an official of the agency and a regular rider will be more likely to be included in the newspaper. Background sheets on issues or events are also helpful to the media in developing their own stories about your system. Routinely provide this information to key media contacts.

If there is no one on the agency staff that has experience in writing stories or announcements in the style used by newspapers, radio stations or television stations, the most convenient and least expensive way to gain this expertise could be contacting a local high school, junior college, or college journalism teacher and borrowing a textbook. There are also some excellent texts available through the mass media that provide the basics of writing and formatting news releases. The following are currently readily available in bookstores, or through online booksellers:

- *Bulletproof News Releases: Help at Last for the Publicity Deficient*, Kay Borden, Franklin Sarrett Publishers, January 1994.
- *Writing Effective News Releases… : How to Get Free Publicity for Yourself, Your Business, or Your Organization*, Catherine V. McIntyre, Piccadilly Books, June 1992.
- *Marketing for Dummies*, Alexander Hiam, IDG Books Worldwide, Inc., 1997.
- *Guerrilla P.R.*, Michael Levine, HarperBusiness, 1993.

Some Spokesperson Do's and Don'ts

Below is a checklist of Do's and Don'ts for the person who is going to be the active spokesperson, or back-up spokesperson for the agency. This list was compiled by a corporate communicator, but is just as appropriate for the public sector.

Some Do's...

DO   Practice the art of bridging - moving a conversation from where someone else wants it to where YOU want it. You do it already in everyday conversation; look for opportunities to improve your skills. It should always be done with finesse and a smile.

DO   Practice the art of turning a negative into a positive. Remember that the very nature of news is what went wrong, and an interviewer's questions will generally, quite naturally, be couched in negative terms. For every negative, there is a positive. Don't be afraid to "bite the bullet," acknowledging the negative, and without pausing for breath, move directly to the positive points you can make.

DO   Remember the value of a smile and a handshake. Even when all else has gone wrong, such positive displays of body language can quite literally save the day.

DO   Do your homework. Your interviewers will have done theirs. Even if the topic is your specialty, do your homework. Don't forget that the interviewer will have gone back to the very basics.

DO   Recognize the value of playing "the Devil's Advocate," even if you must be your own. If at all possible, involve your professional peers. Have them demand that you respond to the toughest questions possible, in private, before you go public. Make sure you have acceptable answers. Anticipate.

DO   Remember your audience. Could the average 10th grade student understand what you are saying? Would he or she care? Are you using language he or she can comprehend? Avoid technical jargon.

DO   Go beyond the interviewer for understanding. If it is worth your time and energy, it should be worth it for you to affect the outcome of understanding.

DO   Speak in 30 second quotes. In spot news situations only, remember, the longer answer cries for editing. The 30 second answer goes as stated in 90 percent of the cases. Eliminate superfluous verbiage. Stay with the facts.

...and some Don'ts

DON'T Talk about things you know nothing about. No "third party" discussions, no answers to hypothetical questions, no "what if" speculation, and no conversation about what some other entity may be thinking, planning or doing.

DON'T Bluff or lie. Morality aside, a good newsperson will know you are lying, expose you for doing so, and your credibility, already low, will be destroyed.

DON'T Be afraid to admit that you don't know the answer to a specific question. If you don't know, say so, and add the magic words, "...but I will find out and let you know." Such an admission, done with candor, can make more of an impression with the viewer than all of your facts and figures combined.

DON'T Give the interviewer ammunition. If you're an oil company executive, it's poor procedure to mention the excellent mileage your wife's Mercedes Benz gets not that you've paid someone to remove the air pollution equipment, which leads us to:

DON'T Go off the record. EVER. Unless you are willing to put your personal career and the best interests of your company in the hands of another person.

DON'T Use negative "buzz words" such as "obscene profits, rip-off, disaster, tragedy, holocaust," etc. If your interviewer uses them, don't repeat them. One exception: The accidental loss of human life is a tragedy by definition. Don't refer to such an event as an "unfortunate incident." EVER!

DON'T Lose your temper or weep on television. Either will assure an unedited interview on the 6:00 and 10:00 news.

DON'T Offer personal opinions. You are your company.

DON'T Use the term, "no comment." EVER! In your viewer's mind, it means you are as guilty as if you had committed a crime and then taken the Fifth Amendment.

DON'T Waste time with preface remarks. Remember the pyramid, and open with the point you want to make. Don't tell what you want to talk about, talk about it!

In summary: Control is the key, achieved by leading, not following, by being excited and enthusiastic about your subject, by having a reason to be there, and that reason obviously, clearly important to you.

Planning for Successful Events

When surveying public transit systems throughout the country for their best low-cost, effective marketing ideas, we found that even the smallest systems, with no stated budget for marketing, were often able to achieve big results with some sort of public event. Whether it is an annual open house for advisory board members and the staff, or major events with substantial corporate sponsorship, there are some common considerations to planning a successful event.

Below are a series of questions that public transit marketing professionals or managers should pose as they plan an event. The questions are based on advice on the subject included in the handbook, Promotions, Publicity and all that Pizazz - Round Two by the Ontario Urban Transit Association's Centre for Transit Improvement."

- First, why are you having an event? What is in it for your transit system? What is in it for your target audience?
- What is the goal of the event? What materials and effort will be required?
- When should you have the event? When will it have the greatest impact (what day of the week, time of day, season)?
- When can you schedule the event to have the least competition from other activities? When does it make the most sense for your own transit system schedule (in terms of competing with staff involvement in service or schedule changes, other major events)?
- When will you need to begin planning to allow for all of the details to be accomplished? (Be sure to leave time to acquire the approvals, sponsorship, suitable notification time to participants.)
- Who will be involved in planning and executing the program or event? Who else is available to assist or participate?
- Who should be considered for possibly co-hosting an event or sponsoring the activity?
- Who will speak at the event? Who will gain you the best advantage for media coverage?
- Who will be the guests/audience at the event? Who is your target audience?
- Where will you hold the event? Where can you get the most for your dollar? (Where can you use a location for no charge?) Where can you ensure appropriate accessibility? Where is a location that meets all of these requirements and is also on a transit route, with available parking, and easy to find?
- Where will the audience or participants be most comfortable?
- How will the program or event be funded?
- How will you determine if the event will be worth the effort? How will taxpayers perceive the event?

CATEGORIES

Categories of Projects/Strategies

For the purposes of this document, public transit system marketing activities have been organized in the categories listed below. Examples of each category of projects or strategies are provided in the sections that follow.

Accessibility-Related Projects

Transportation services for the disabled are an important part of the transit agency's story to communicate both to the target community and the general public. Many individuals with disabilities are provided with accessibility through public transit services that is imperative to their daily lives. Public information campaigns about the accessibility services provide instructions for using the service, and also serve to enhance the agency's image as a vital part of the community.
(A-1 to A-3)

Community Events

Community events provide the opportunity to demonstrate the place a transit agency has within the fabric of a community. The populace gains a sense of confidence in its local government when a public agency shows commitment to an important local event. It is also an opportunity to forge good relations with other important community organizations.
(B-1 to B-6)

Cooperative Promotions

Cooperative promotions with the private sector have multiple pay-offs for the public transit system. While the savings realized by the public transit system in sharing the costs of a marketing project with the private sector may be the first pay-off that comes to mind, the development of goodwill with the private sector can also become a long-term benefit. The company that joins the transit system in a promotional effort may become a participant in the regular subsidized employee pass program, for example. The image conveyed to the public by coupling the transit system with an established retail name may enhance the general perception of the transit system. Projects or strategies included in this section include those where there were services contributed by the private sector; those where money was contributed by the private sector for the execution of the project; or those that were joint promotions.
(C-1 to C-15)

Image Promotions

Image promotion builds connections between the transit agency and the community. The goal of an image promotion is to present a positive image of the transit industry in general and the agency. Unlike a promotion about a specific service or route, image promotions are often targeting the community at large. It is a good opportunity to present the benefits of the public subsidized transit system to the community.
(D-1 to D-9)

Internal Promotions

Internal promotions are an opportunity for a transit agency to enhance the organization from within, with the attendant result of strengthening the level of service provided to riders. The level of morale within a transit organization directly affects its level of customer service. Good internal promotions are effective morale boosters. They enhance the unity and confidence of staff and build good two-way communication between management and employees. Internal promotions can respond to both short- and long-term agency needs, all the while building credibility with the community and creating a positive public image.
(E-1 to E-5)

Introduction of New Service

The introduction of new service is a time for celebration at a transit

agency, both internally, and if the change is large enough, publicly. The agency is growing and improving to meet new customer demands and is seen as responding to the city's development. The agency is evolving with its market. There are several ways to promote new services and many purposes that can be served. A strong promotion can get a new route or service off to a vigorous start. Local businesses that are positively affected by the new service are often willing to cooperate in promoting the service. Promotions can also reward current riders and attract new ones.
(F-1 to F-3)

Media Relations

The purpose of good media relations is to promote understanding, goodwill, and acceptance of transit by the public. Utilizing local media to promote transit events or news is a very effective method of disseminating information. Establishing solid lines of communication with local media representatives can assure more fair and accurate coverage when the agency faces a crisis or has important transit-related news occurs to ensure fair and informed media coverage.
(G-1 to G-5)

Problem-Solving Projects

On occasion, problems arise that a transit agency may address through marketing efforts. The problem may be one of perception by the public or internally by employees. There could also be situations that inhibit an agency from performing at its maximum level of service. A transit agency that is seen taking concrete steps to address problems as they occur, and letting people know it, retains the confidence of passengers, employees, local government, and the general public.
(H-1 to H-8)

Promoting Transit

Promoting transit as a viable option in the mix of transportation alternatives is essential to the success of a transit agency. Citizens in the agency's service area may be unaware of the convenience of using transit for their daily activities. Creative marketing campaigns will enhance the perception of transit as an effective alternative form of transportation and perhaps lead to new riders who may be unaware of its availability or advantages of using it. Promoting the transit service as a worthwhile public service is also helpful in attracting and maintaining support from

non-users who may support that service with their local, state, and/or federal taxes.
(I-1 to I-9)

Rider Inducements

Transit agencies seek the most effective ways to induce individuals to become regular users of a transit system. Just as important as new riders, the agency must find ways to reward its regular customers. Adding value to the transit experience of core ridership helps to maintain a loyal customer base.
(J-1 to J-10)

Seasonal Promotions

Seasonal activities offer an excellent opportunity to promote a transit agency and the services it provides. Few other times of the year can create and attract as much transit ridership as holidays. As a result, the transit system is on display. Events that attract a large number of citizens create a demand for the use of transit, especially if held in areas that contain limited access or parking. The community perceives increased use of the transit agency as lessening the potential for accidents and injuries since fewer drivers are in their own cars. By providing increased public safety during holidays, especially ones traditionally cele-

brated with alcohol such as New Year's Eve, the transit agency is seen as providing a public good and its image is enhanced. A well-organized, efficient transit program during a holiday season or weekend can reap many rewards for an agency.
(K-1 to K-3)

Special Events

Special events enable transit systems to participate in community efforts or to provide transit services to community activities. Transit systems may also choose to create events to familiarize infrequent transit riders with available services. Special events may provide good anchors for other targeted activities. The special event allows personal contact with a large number of people in a concentrated time period and is invaluable in launching or culminating advertising or promotional campaigns. State and county fairs provide the same marketing opportunities as any special event does. Every state has a fair, often requiring effective transit services for success. For transit systems not located near the annual state fair, the same ideas and efforts can be applied to county and regional fairs.
(L-1 to L-6)

Target Group Promotions

Many groups can be targeted in promotions by a transit agency to boost ridership, to reward current users of the system, to entice new ridership, and to educate segments of the population in the use and value of public transportation. Groups that are included in the examples below include: senior citizens, new employees, school children, high schools, users of libraries, and residents adjacent to bus routes. *(M-1 to M-25)*

Try Transit Week

Try Transit Week is an annual event initiated by the American Public Transit Association. The event is observed by public transit systems across the nation with a variety of activities, such as a day of reduced fares or no fares at some systems, educational presentations at schools and civic organizations, and distribution of transit marketing materials. Try Transit Week helps identify the benefits of using public transportation to current and potential riders, and often showcases a transit system's most creative marketing ideas. *(N-1 to N-6)*

Accessibility-Related Projects

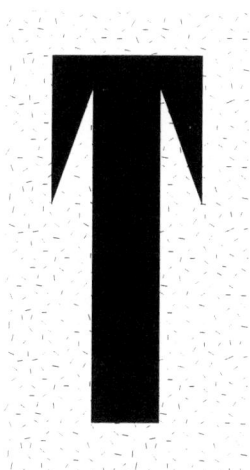

ransportation services for the disabled are an important part of the transit agency's story to communicate both to the target community and the general public. Many individuals with disabilities are provided with accessibility through public transit services that is imperative to their daily lives. Public information campaigns about the accessibility services provide instructions for using the service, and also serve to enhance the agency's image as a vital part of the community.

Radio Ad for Rider Training Program for People with Disabilities (A-1)
Capital Metro

Number of Vehicles: 303 buses, 20 trolleys, 76 special transit vehicles

Strategy
"Paul Hunt, Walk in My Shoes" is the title of a 60-second radio ad promoting a free rider training program by Capital Metropolitan Transportation Authority of Austin, TX for citizens with disabilities. The ad allows listeners the opportunity to experience the challenges and concerns of a visually impaired person negotiating city streets. Listeners are taken step-by-step as Mr. Hunt attempts to cross a busy intersection. It also informs people with disabilities that Capital Metro provides free training to visually and mobility-impaired citizens in order to use the bus system for daily activities.

Objectives
To increase awareness of an additional service Capital Metro provides to individuals with disabilities.

Resources
The radio spot was produced under contract by the Austin advertising and public relations firm of TateAustin. The ad cost approximately $500 and required 30 hours of staff time. Paul Hunt is past chairman of the agency's Mobility Impaired Services Advisory Committee and volunteered to do the ad.

Implementation Time
Three weeks

Results
The radio ad received a Barbara Jordan Award medallion in 1996 by the Texas Governor's Committee on People with Disabilities, which recognizes outstanding media efforts that encourage accurate and progressive portrayals of persons with disabilities. Judges noted its "refreshing and innovative approach." While involvement in the training program remained at the same level, Capital Metro did notice an increase in awareness of the service based on phone calls and referrals from community agencies and persons with disabilities interested in the training program.

Adaptations
The tape of the ad was used in disability awareness training for transportation personnel and training for effective media approaches to enhance services to persons with disabilities.

When
Created in 1995, broadcast in 1996

Contact
Nancy D. Crowther
Accessible Transportation Specialist
Capital Metro Transportation Authority
2910 East 5th St.
Austin, TX 78702
Tel: 512/389-7583
Fax: 512/369-6072
E-mail: nancy.crowther@capmetro.austin.tx.us

TRIP Center Directory (A-2)   *Wheels, Inc.*

Number of Vehicles: 150 (brokered)

Strategy

The TRIP (Transportation Referral and Information Program) Center Directory is a resource offered by WHEELS, Inc. that lists 260 sources of passenger transportation available in the Greater Philadelphia area. Each entity listed supplies special transportation needs, either in whole or in part, to persons with special transportation needs. The service offered by each is described in a one-page profile. The profile is in an outline format which identifies the agency and provides details regarding its operations, passenger characteristics, methods by which transport is obtained, and fare requirement. It also cross-references the names of other transportation services with which the agency is involved. The profile is held in a data base at WHEELS, Inc. which is regularly updated. The directory is distributed free of charge to area public libraries and major human service agencies, and is sold throughout the region for $87.00 ($80.00 for non-profit organizations).

Objectives

To provide a comprehensive source of information on area services that serve special transportation needs.

To maintain the specialized transportation leadership reputation of WHEELS, Inc.

Resources

Financing for the project was provided by extensive foundation support, WHEELS, Inc. program funds, and paid advertising within the directory. The total cost for the project was $35,000. The project required four professional staff members dedicating several hours per week in the course of a year. Six part-time, temporary personnel were utilized to conduct surveys to acquire data on the agencies. A consultant was hired to develop the database at a cost of $5,000. A professional designer and printer was used for the final publication of the directory at a cost of $3,000.

Implementation Time

One year

Results

The directory was published in 1995 and remains the only source of its type of information available to the public. A TRIP Center office was established. However, only 200 copies of the directory were purchased from an original printing of 800 copies.

Adaptations

It is the desire of WHEELS, Inc. to make the directory more interactive to facilitate updates, and perhaps make the directory available through computer media.

When

Initiated in 1994 and ongoing

Contact

Greg Ficchi
Data/Automation Manager
WHEELS, Inc.
1118 Market St.
Philadelphia, PA 19107
Tel: 215/563-2000
Fax: 215/563-5531
E-mail: wheels @libertynet.org

Promoting Fixed-Route Service for Medical Trips (A-3)
WHEELS, Inc.

Number of Vehicles: 150 (brokered)

Strategy

In Philadelphia, Medical Assistance clients of the Pennsylvania Department of Public Welfare who are eligible for non-emergency trips to outpatient medical services must register with and be screened by WHEELS, Inc. It was determined by WHEELS, Inc. that many clients were physically and developmentally able to make these medical trips using Southeastern Philadelphia Transportation Authority (SEPTA), the local public transit system. The screening process results in the able clients being "assigned" to use of SEPTA and prevents them from using the program's paratransit service. WHEELS, Inc. reimburses the clients for their out-of-pocket expenses up to the cost of a monthly transit pass, issuing checks monthly in time to purchase their fares for the coming month. Fare reimbursement claims are validated by the medical provider and WHEELS, Inc. staff. Complete instructions and materials are supplied to the clients and medical providers.

Objectives

To utilize existing transportation services, thereby reducing program costs while ensuring satisfaction of trip needs. To reduce the volume of paratransit services supplied by the program.

Resources

Funding for the project was provided by the Medical Assistance Transportation Program of the Pennsylvania Department of Public Welfare. Twenty percent of the internal staff employees were utilized.

Implementation Time

The concept was initiated upon assumption of the project by WHEELS, Inc.

Results

Before this technique was utilized, the unit cost for program trips was up to $20, all via some form of paratransit. Using public transit, the cost to the program is $1 to $2 per trip. Almost one-half of all program trips are now made on SEPTA with no loss in fulfilling client trip requirements.

Adaptations

WHEELS, Inc., uses a similar screening process for a Title 1 project.

When

1983 and continuing.

Contact

Ross Dougherty
Controller
WHEELS, Inc.
1118 Market St.
Philadelphia, PA 19107
Tel: 215/563-2000 x220
Fax: 215/563-5531
E-mail:
wheels@libertynet.org

Community
Events

Community
Events

ommunity events provide the opportunity to demonstrate the place a transit agency has within the fabric of a community. The populace gains a sense of confidence in its local government when a public agency shows commitment to an important local event. It is also an opportunity to forge good relations with other important community organizations.

Co-sponsor of a Clean Air Fair (B-1) *Transit Authority of Northern Kentucky*

Number of Vehicles: 107 buses

Strategy
The Transit Authority of Northern Kentucky (TANK) of Fort Wright, KY co-sponsored a Clean Air Fair and a 3.5-mile "Stride to Breathe" Walk. The event was held in order to raise awareness of clean air issues within the community. It also promoted mass transit as a viable transportation alternative.

A local committee, Northern Kentuckians for Cleaner Air, organized the walk and the fair. The events were publicized through public service announcements on local radio stations and in newspaper ads. Flyers were also distributed at strategic points around the community. A TANK bus, driven by the agency's mascot, led the walk and passed out water to participants.

The fair provided booths from the local health department, the transit agency, the American Lung Association, the Department for Environmental Protection's Division of Air Quality, and offered allergy screenings, food and entertainment.

Objectives
To raise awareness of clean air issues in the community.

To get people to view mass transit as a viable transportation alternative.

Resources
Funding for the event came from the Regional Ozone Coalition (ROC), which covers an eight-county region. The ROC is part of the Ohio-Kentucky-Indiana Regional Council of Governments. Total cost of the project was $5,000. Two TANK staff members worked on the event for a total of 80 labor-hours.

Implementation Time
Four months

Results
The "Stride to Breathe" Walk saw mixed success due to poor weather conditions. Relations between the agency and the ROC were enhanced.

When
May 1997

Contact
Gina Shipley
Manager of Marketing and Planning
Transit Authority of Northern Kentucky
3375 Madison Pike
Fort Wright, KY 41017-9670
Tel: 606/331-8265
Fax: 606/578-6952
E-mail: gshipley@fuse.net

Person of the Year with Disabilities Award (B-2)
Good Wheels, Inc.

Number of Vehicles: 72 buses

Strategy
To highlight the good works of its clients, Good Wheels, Inc. of Fort Myers, FL sponsors a "Person of the Year with Disabilities" Award. Nominations can be made by anyone in Southwest Florida. No one in the market area had developed such an award previously.

Good Wheels used its public relations firm to organize the event. As part of the research, the public relations firm met with representatives from eight other social service organizations that serve the disabled and disadvantaged population.

The first step was to obtain sponsorships. Packets were sent to area hospitals, health insurance companies, and other organizations that serve individuals with disabilities. Follow-up phone calls and visits were made to potential sponsors. Nomination forms were distributed to more than 100 social service agencies in the region. Media packets were sent and radio and television interviews were scheduled. A panel of judges was selected from community leaders (including one individual with disabilities) and was moderated by the chairman of the Good Wheels Board of Directors. Good Wheels provided free transportation to the luncheon for any individual with disabilities requesting service.

Objectives
To present the Person of the Year With Disabilities Award and recognize five finalists.

To obtain positive pre- and post-award local media coverage of the nomination process and the awards luncheon.

To attract a minimum of 100 people and several dignitaries to the awards luncheon.

To obtain $2,500 in sponsorships to keep ticket prices low enough to avoid an event subsidy by Good Wheels.

Resources
All expenses were paid with revenues from sponsorship fees of $2,200 and ticket sales of $1,200.

Implementation Time
Three months per year.

Results
Three rounds of media packets produced a great deal of publicity for the nomination process, the judging, and the awards luncheon. In each case, the event received coverage from the daily newspaper, weekly newspapers, all four television network affiliates, and radio stations. Altogether, there were 154 inches of print media, 8 minutes of television broadcast time, and 3 hours of radio public service time devoted to the event.

Eight sponsorships were obtained, totaling $2,200. Ticket prices for those attending the awards luncheon (nominees attended at no charge) were held to $15 per person.

The award winner was chosen from a field of 30 nominations. Five finalists were also recognized. A total of 125 individuals attended the awards luncheon and it attracted state legislators, city council members, and county commissioners.

The event was recognized in 1996 by the Florida Public Relations Association as one of the finest special events of the year. It also won a Judge's Award for cost-effectiveness.

When
1995 and awarded annually

Contact
Deloris Sheridan
President
Good Wheels, Inc.
10075 Bavaria Rd.SE
Fort Myers, Fl33913
Tel: 941/768-6184
Fax: 941/768-6187
E-mail: gowheel@aol.com

Clean Air Challenge (B-3)
Sacramento Regional Transit District

Number of Vehicles: 210 buses, 36 light rail vehicles

Strategy
In conjunction with the American Lung Association of Sacramento Emigrant Trails, Sacramento Regional Transit challenged two area high schools to clean the air and clear the school parking lots of automobiles by riding transit on May 13, 1998, as part of Clean Air Month and Try Transit Week activities.

The transit agency provided schools with bus schedules, maps, and safety brochures. Agency staff visited both schools prior to the event to display transit information, answer questions, and provide "how-to" route information. "New Rider" complimentary tickets were also offered.

Objectives
To promote alternative transportation to teenagers for the day and plant a seed for future use of alternative transportation.

Resources
No formal budget was required for the project. Existing materials, schedules, maps, and displays were used. One marketing representative and six customer service representatives were utilized.

Implementation Time
Three months

Results
The goal of a 25 percent reduction in automobile use for the day was met at both schools. More than 2,300 students carpooled, biked, walked, or took the bus.

When
May 1997 and repeated annually.

Contact
Jo Teele Noble
Marketing Representative
Sacramento Regional
Transit District
PO Box 2110
Sacramento, CA 95812
Tel: 916/321-2863
Fax: 916/444-0502

"Bike the Bay" Campaign (B-4)
Hillsborough Area Regional Transit Authority

Number of Vehicles: 176 buses

Strategy
As part of a promotion for the Bikes on Buses program, Hillsborough Area Regional Transit Authority (HARTLine) of Tampa, FL created the "Bike the Bay" Campaign. The community event included two group bike rides from HARTLine's downtown commuter center, one short and the other longer. The shorter distance is 10 miles, the longer ride is 23 miles. Participation is free and the event is held once a year on a Saturday to allow for maximum participation.

HARTLine promoted the campaign through various means: newspaper and radio ads, press releases, bus public service announcements, patron newsletter information, outreach to local bike shops and colleges, and direct mail to the agency's Bikes on Buses database (all users of HARTLine's bike racks must fill-out a form and get a picture ID).

Objectives
To promote the Bikes on Buses program as a transit alternative.

Resources
The event requires 40 hours of HARTLine staff time. The Tampa Tribune, a local daily newspaper, donated $5,000 in advertisements. Food and drinks were donated by vendors. Radio ads cost $1,500, while promotional materials totaled $2,000.

Implementation Time
Two weeks

Results
The first year of the event saw 42 bicycle riders participate, while the second year attracted 146 cyclists. Bikes on Buses program usage increased 98.6 percent during the second year of the event.

When
1996 and conducted annually

Contact
Pat McElroy
Promotions and Advertising Specialist
Hillsborough Area
Regional Transit Authority
4305 E. 21st Av.
Tampa, FL33605
Tel: 813/223-6831
Fax: 813/223-7976

"Our Own Words" Poetry Contest (B-5)   *Pierce Transit*

Number of Vehicles: 193 buses, 226 vans

Strategy
In order to partner with other organizations and businesses in a meaningful community relations effort, Pierce Transit of Tacoma, WA developed "Our Own Words," an annual teen poetry and fiction writing contest, and received the co-sponsorship of the Pierce County Library Foundation, the Tacoma Public Library, and the Puget Sound Poetry Connection. Besides cash and other prizes, winners have their writing excerpted on colorful exterior bus boards appearing on Pierce Transit buses throughout the month of May.

Joint planning with the co-sponsors begins in September. Budget, timeline, and judging procedures are established. Prizes and promotional agreements are acquired. Marketing materials, such as flyers and posters, to be distributed to schools, libraries, and transit centers, are created by late January. Contest packets are then mailed to teachers and notices and entry forms are posted on the Pierce Transit website. Meetings are held with school district and art association representatives to enlist their help in promoting the contest.

A detailed media release is distributed before and after the contest to generate interest. Five newspaper ads are run promoting the contest, and one is displayed after the award ceremony congratulating winners. Follow-up letters are sent to all school principals notifying them of the number of students who entered the contest. An awards ceremony is held in early May at which the winners read from their work and receive their prizes, including a commemorative book of all the winning entries. Twenty exterior bus boards featuring winning entries travel throughout the county in May.

Objectives
To partner with other organizations and businesses on a meaningful community relations effort that generates significant local media coverage and enhances Pierce Transit's reputation for community involvement.

Resources
The cost to Pierce Transit for the project is approximately $5,000. Free advertising for the contest is donated by the Tacoma News-Tribune, the local daily newspaper. Gift certificates used as prizes are provided by Borders Books. The agency was provided with discounted printing of contest materials by a local merchant, and received a discount on facility rental for the prize ceremony.

Implementation Time
Five Months

Results
The contest garnered 609 entries from 82 schools. The awards ceremony drew 100 people, mostly parents, teachers, and judges. Nineteen articles appeared in the local press, most of them with photos of the prize winners standing in front of "their" buses. Many of the winning teens and their families came to Pierce Transit headquarters to take photos of "their" bus. The agency established positive relationships with its co-sponsors.

When
1997 and conducted annually

Contact
Jean Jackman
Public Information Officer
Pierce Transit
PO Box 99070
Tacoma, WA98499-0070
Tel: 253/581-8034
Fax: 253/581-8075
E-mail: jackmanj @piercetransit.org

Stuff-A-Bus Promotion (B-6) *Valley Metro*

Number of Vehicles: 325 buses

Strategy

The "Stuff-A-Bus" campaign provides an opportunity for the public to donate nonperishable food stuffs for those in need. The promotion joins the City of Phoenix Public Transit Department (Valley Metro) with KDKB, a local radio station, and a grocery store chain. KDKB promotes the event while the grocery store chain provides locations to park the bus and receive donations. The donated food goes to a local food bank to help feed low income families in the community.

The event takes place during two weeks in November. Several locations of the sponsoring grocery store chain are used to host the event. The schedule of times and locations are publicized on the radio station so that citizens could plan ahead and visit the closest location. An article about the promotion is featured in the passenger newsletter and press releases are distributed to all media outlets. An article is also featured in the food bank's newsletter. Valley Metro provides volunteers to drive and stay with the bus for the duration of the promotion.

Objectives

To promote goodwill within the community in the face of service cuts and fare increases.

Resources

The only direct cost to Valley Metro is postage for mailing press releases. Two buses are utilized for the promotion. All publicity is donated by KDKB.

Implementation Time

Two months

Results

Over 10,000 pounds of nonperishable food was collected in two buses. Three local television stations covered the unloading of the buses at the food bank.

When

November 1995 and repeated annually

Contact

Angie Harvey
Acting Customer Communications Manager
Valley Metro
302 N. 1st Av. #640
Phoenix, AZ 85003
Tel: 602/261-8255
Fax: 602/261-8756
E-mail: aharvey
@vm.maricopa.gov

Cooperative Promotions

Cooperative Promotions

ooperative promotions with the private sector have multiple pay-offs for the public transit system. While the savings realized by the public transit system in sharing the costs of a marketing project with the private sector may be the first pay-off that comes to mind, the development of goodwill with the private sector can also become a long-term benefit. The company that joins the transit system in a promotional effort may become a participant in the regular subsidized employee pass program, for example. The image conveyed to the public by coupling the transit system with an established retail name may enhance the general perception of the transit system. Projects or strategies included in this section include those where there were services contributed by the private sector; those where money was contributed by the private sector for the execution of the project; or those that were joint promotions.

"Here's the Scoop" Campaign (C-1)
Bloomington Transit

Number of Vehicles: 20 buses

Strategy
Bloomington Transit of Bloomington, IN offered a special promotion called "Here's the Scoop" in order to address a decline in ridership on Saturdays over the course of the previous year. The campaign lowered cash fares from 75 cents to 10 cents on each Saturday in the month of July. Riders also received a coupon from a local ice cream store for a free small ice cream. In return for its participation, the name of the ice cream store was included on all printed advertising for the campaign, as well as in live media coverage.

Objectives
To reverse the decline in ridership on Saturdays, particularly during the summer.

Resources
The total cost of the promotion was approximately $1,300, two-thirds in lost Saturday fares. The ice cream store provided 500 coupons and $250 in advertising expenses.

Implementation Time
One month

Results
The transit agency estimated a 20-25 percent increase in Saturday ridership during the month of the campaign.

When
July 1997

Contact
Polly Freyman
Special Services Coordinator
Bloomington Transit
130 W. Grimes Lane
Bloomington, IN 47403
Tel: 812/332-5688
Fax: 812/332-3660

The Talking Yellow Pages (C-2)   *Oahu Transit Services*

Number of Vehicles: 525 buses

Strategy

The "talking yellow pages" is a service offered by GTE Directories to its customers who purchase ads in the telephone yellow pages. Oahu Transit Services took advantage of the program to offer information about its routes most used by visitors to the island and about its express services. The transit system swapped the cost of the information service for promotion of GTE through in-bus advertising. In addition, the transit system promotes the service (and thereby GTE) in other ways such as brochures, pamphlets, timetables, concierge reference books in most Waikiki hotels, and a 12-minute video shown on the Japanese Visitor Cable Network every two hours in 30,000 Waikiki hotel rooms.

Objectives

To provide 24-hour recorded information service to visitors (35,000 riders daily) and express bus riders (45,000 daily) at virtually no cost to the public transit system.

Resources

An information specialist with the transit agency spends time to provide an information "script" to the telephone company, which is updated as schedule changes occur. The agency estimates that it receives approximately $70,000 annually in GTE services in exchange for $15,000 worth of interior bus advertising. In addition, the City of Honolulu prints brochures also promoting the GTE information system. A local bank (First Hawaiian Bank) also underwrites system costs.

Implementation Time

Six months

Results

The use of the computerized telephone information system is recorded daily. Current numbers (1997) show more than 200 calls from visitors daily and more than 100 calls about express bus services.

Adaptations

GTE and other telephone directory providers have similar services throughout the United States. Oahu Transit Services believes it is the first to use the program for bus schedule information. The success of this Hawaiian program may be used as a model for transit systems large or small to pursue similar cooperative ventures.

When

Initiated in 1993 and ongoing

Contact

William L. Haig
Manager of Customer Services
Oahu Transit Services (The Bus)
811 Middle St.
Honolulu, Hawaii 96819-2388
Tel: 808/848-4501
Fax: 808/848-4419
E-mail: bhaig@thebus.org

Rider's Guide Publication (C-3)
Kent State University Campus Bus Service

Number of Vehicles: 23 buses

Strategy

The Rider's Guide is a booklet with information and maps about Kent State Campus Transportation local bus routes including Kent State University and Kent, OH. The booklet is given to all incoming freshman, distributed throughout campus, and given to the communities around Kent.

As part of a campaign to offset some of the costs of production, advertising in the guide is sold to local businesses. In return for their purchase of ads, the business name is printed on the route maps to indicate location of the establishments. A list of advertisers and phone numbers is included at the back of the directory.

The guide is a time critical publication. There are several deadlines and, as a result, early solicitation of businesses and organizations is essential to success. Good communication between the marketing and graphics departments, along with the printer, are also important.

Objectives

To garner sponsorship to offset publication costs of the transit guide.

Resources

The Rider's Guide is an annual publication, requiring approximately 150 hours of staff time to produce. The cost of the publication varies annually.

Implementation Time

Approximately five months.

Results

In 1996-1997, all advertisement space in the guide was completely sold for the first time. The agency was able to cover all production costs through advertising sales for the first time in 1997-1998.

When

The guide is published annually.

Contact

Elizabeth Ricchiuti
Marketing Associate
Kent State University Bus Service
1950 SR 59
Kent, OH 44240
Tel: 330/672-7433
Fax: 330/672-3662
E-mail:
ericchiuti@kent.edu

RIDER'S GUIDE

Kent State University

1996-97

CAMPUS BUS SERVICE 672-RIDE

672-RIDE

CAMPUSBUSSERVICE
GO WITH IT!

Co-op Program with the Knoxville Museum of Art (C-4)
Knoxville Area Transit

Number of Vehicles: 80 buses, 20 lift vans

Strategy
Knoxville Area Transit (KAT) of Knoxville, TN negotiated a cooperative promotion with the Knoxville Museum of Art (KMA) for a summer exhibit by the artist Red Grooms. KAT provided a wrapped bus painted red with the exhibit logo for display at the museum on certain days during the exhibit. The agency also used the bus for shuttles, for excursions, as a rolling billboard, and as a display at other events. In return, KMA provided the transit agency with recognition as a sponsor in printed brochures, invitations, newsletters, and media advertising.

Objectives
To increase awareness of KAT and mass transit in Knoxville.

To increase attendance at the museum exhibit.

Resources
The agency and the museum, by written agreement, evenly traded $2,400 worth of services with each other.

Implementation Time
One month

Results
The agreement worked to the mutual benefit of both organizations. Many people took note of the brilliant red-wrapped bus and positive comments were heard about the transit system.

Adaptations
Other groups within the city have sought mutually beneficial agreements with KAT.

When
May-August 1997

Contact
Belinda Woodiel
Marketing Manager
Knoxville Area Transit
1135 Magnolia Av.
Knoxville, TN 37917-7740
Tel: 423/546-3752
Fax: 423/525-5240

"Sally Says" Placemats for McDonald's (C-5)
Citizens Area Transit

Number of Vehicles: 192 buses, 104 paratransit vehicles

Strategy

"Sally" is a cartoon character whose name represents "save a life like yours." The transit system was authorized by the state's office of traffic safety - bicycle and pedestrian program to use the character in combination with McDonald's Corporation cartoon characters (Ronald McDonald and friends) in developing a printed piece in a cooperative venture with McDonald's. The fast food company distributed the printed piece (Colorful tray placemat) by alternating the placemat with its other promotional placemats during a 12-18 month period. A kick-off program for the use of the placemat including a media event featuring elected officials, the private sector co-sponsors, and children. The state traffic safety program also provided a costume (usually worn by a 12-year old girl) to be used in public appearances promoting the program.

Objectives

To increase public awareness regarding pedestrian safety, specifically the safety of pedestrian children.

Resources

The major expense of the strategy was printing. The agency spent $8,000 for the placemat printing. Other estimated costs include time spent by agency staff on the project: marketing manager - 10 hours; graphic artist - 40 hours; and staffing of the kick-off event - 4 hours. Underwriting much of the agency's cost of the project, however, was a state transportation safety grant.

Implementation Time

Ninety days

Results

The placemats were widely distributed at area McDonald's. They were also used by elementary school teachers as a classroom tool for discussing safety. While there is no way of attributing a reduction of pedestrian accidents directly to the placemats, the Regional Transportation Commission (RTC) staff judged the program successful in promoting Citizens Area Transit (CAT) as a good citizen. It is conceivable that the cooperative project also led to the involvement of McDonald's in the new CAT downtown transfer center. Another cooperative program - production of a plastic cup to commemorate the opening of the transit center - was undertaken at that time.

Adaptations

Because of the success with the McDonald's placemat, the RTC is considering other similar joint endeavors to reach different audiences.

When

1996-1997

Contact

Stanton Wilkerson
Marketing Manager
Regional Transportation
Commission of Clark County
301 East Clark St., Suite 300
Las Vegas, NV 89101
Tel: 702/455-5940
Fax: 702/455-5959
E-mail: rtc@co.clark.nv.us

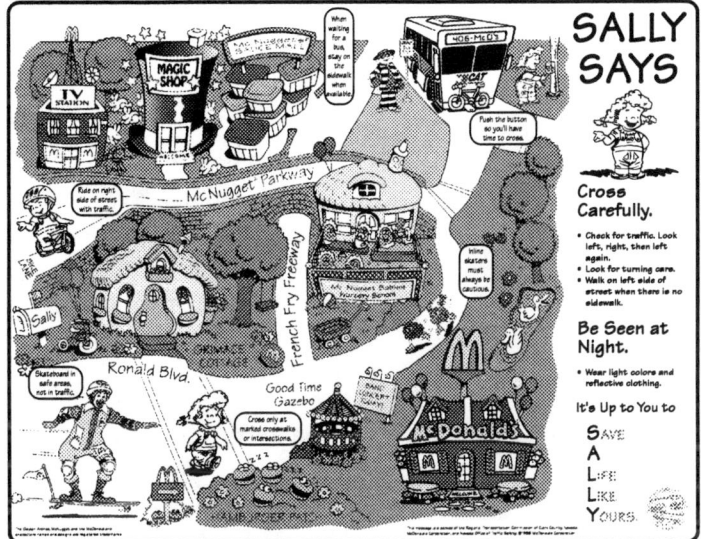

New Residents Program (C-6)

Tri-County Metropolitan Transportation District of Oregon

Number of Vehicles: 766 buses, 26 light rail vehicles

Strategy

In January 1989, the Tri-County Metropolitan District of Oregon (Tri-Met) in Portland launched a monthly marketing program directed towards new residents in the Tri-Met district. New residents in 24 zip codes within the Tri-Met district were targeted for the promotion. Target areas were chosen because of particularly good transit service. CPC, a Pennsylvania firm that specializes in new resident promotions, was hired to obtain the names and addresses of new residents in the target zip codes and to mail a promotional packet each month. In June 1994, Tri-Met arranged for Portland General Electric (PGE), a local power company, to mail a packet to new residents within two weeks of the move-in date. CPC continues to mail the packet to areas not served by PGE.

The packet contains a letter outlining the personal benefits of riding transit, a coupon that can be redeemed for a packet of information about riding Tri-Met, and three free day tickets. One-half of the packet is an offer to plan a trip of the respondent's choosing on the system. A short survey appears on the back of the response coupon which obtains general information about the respondent's riding behavior to enable the agency to more carefully segment its target market.

Objectives

To increase ridership on Tri-Met by targeting new residents and informing them of the service the agency provides.

Resources

The cost of maintaining the program is approximately $35,000 per year. Minimal staff time is required to maintain the program.

Implementation Time

Six months

Results

The program was considered very successful by Tri-Met. The free tickets and response rate to the survey exceeded the agency's expectations. The success of the promotion demonstrates that moving is a prime time to effect behavioral changes such as switching modes of transportation. The keys to the promotion's success were carefully segmenting out Tri-Met's target market and providing complete information to citizens at a time when they are making major changes.

When

Program began in 1989 and is continuing.

Contact

Rogene Clements
Administrative Specialist
Tri-Met
4012 Southeast 17th Ave.
Portland, OR 97202-3993
Tel: 503/238-4917
Fax: 503/239-6469

Co-op Grocery Store Program for Seniors (C-7)
Beaver County Transit Authority

Number of Vehicles: 42 buses

Strategy
In early 1993, the only grocery store within walking distance of senior citizen housing complexes in Beaver Falls, PA closed. Of immediate concern was how to ensure the affected citizens, primarily the elderly, could conveniently purchase groceries. An ad hoc committee of Beaver County Transit Authority (BCTA) staff, Beaver County Office on Aging (BCOA), senior citizen complex managers, Housing Authority staff, and several grocery store managers was formed to study the situation and find possible solutions. The committee proposed a demand-responsive, door-to-door shuttle service between the senior citizen complexes and the remaining grocery stores. Currently, twelve complexes have programs with five different stores. Approximately 275 one-way passenger trips occur each week between the complexes and the stores. Each shopping trip lasts two hours. Senior citizens in the complexes are encouraged to reserve trips at least 24 hours in advance, preferably one week in advance, before the trip. The service is funded by the Pennsylvania Lottery Program (85 percent) and BCOA (15 percent), with the customer paying fifty cents towards each trip. On the inaugural trip to the grocery store, each participant is given an identification name badge with a designated number. This expedites recognition of groceries after purchasing. A matching ID number is placed on each bag of groceries. The participants are encouraged to stop at the store's conference room for snacks courtesy of the store. The grocery store delivers the bags to the community room of each complex, which eliminates the need for seniors to negotiate handling the bags on the bus on the return trip.

Objectives
To help senior citizens feel independent and remain part of the community.

To transport senior citizens to the grocery store on a scheduled basis.

To induce ridership.

Resources
The primary requirement for the program was a comprehensive meeting with all organizations to coordinate efforts. The program is funded through the Pennsylvania Lottery Program and BCOA.

Implementation Time
One month

Results
The program began with just three senior citizen apartment complexes and one grocery store. It now serves 12 complexes and 5 participating stores. As news of the program is disseminated in the community, additional complex managers call for information about the service.

When
1993 and continuing

Contact
Rose Sutter
Paratransit Manager
Beaver County Transit Authority
200 W. Washington St.
Rochester, PA 15074
Tel: 724/728-4255
Fax: 724/728-8333

52

Bus Book Advertising (C-8)
Citifare/Regional Transportation Commission

Number of Vehicles: 95 buses

Strategy
In order to minimize litter on buses and in transfer stations, and to provide a consolidated transit system information source, Citifare of Reno, NV created a comprehensive Bus Book that contains all routes, maps, and rider information. In order to offset production costs, Citifare sold advertising in the booklet to local companies and merchants.

Citifare targeted current riders rather than the general public as the market for the Bus Book. It used exterior queen panels, interior panels, small posters at transfer stations, as well as ads placed in the regular print media schedule (apartment guides, senior citizen newsletters) to promote the new Bus Book. A simple message showed "before" and "after" photos with models holding all of the individual timetables versus the one single bus book.

Objectives
To develop a better product for passengers that puts all routes, maps, and rider information in one easy-to-use booklet and is subsidized by advertising sales.

To reduce the litter on Citifare buses and in transfer stations created by discarded individual timetables.

Resources
The total promotion budget for the Bus Book was $15,250. This included advertising and development, advertising printing, media and public relations.

Implementation Time
The Bus Book is a continuous project.

Results
Letters, phone calls, and letters to the editor indicate a high level of satisfaction with the booklet. Litter on buses and at transfer stations was reduced considerably.

When
March 1997 and updated twice a year

Contact
Terry McCloud
Ad Sales Specialist
Regional Transportation
Commission/Citifare
PO Box 30002
Reno, Nevada 89520
Tel: 702/348-0400
Fax: 702/324-3503
E-mail: tmccloud
@rtcwashoe.com

Painted Bus Program (C-9)   *Rockford Mass Transit District*

Number of Vehicles: 37 buses, 1 trolley, 18 para-transit

Strategy
In order to ease the difficulty in attending the city's summer festival, On the Waterfront, the Rockford Mass Transit District (RMTD) of Rockford, IL, in conjunction with local sponsors, provides a free Park-n-Ride service to the festival. The festival is held in the downtown district where parking is limited. The sponsor's painted bus was used to provide the service. The sponsors include a local television station, a local merchant, and a pork producer's association. RMTD worked in conjunction with the festival sponsors and promoted the service through the local media.

Objectives
To provide quick and easy access to the festival for those attending from out of town.

Resources
The cost of the service is covered by the sponsors. The agency's expenditures are approximately $550 in time and fuel.

Implementation Time
One week

Results
Five hundred riders were anticipated, the final count was 2,600. RMTD ended up requiring two buses for the service, when only one had originally been foreseen.

Adaptations
The agency provides shuttle service with its painted buses for a pro-am golf tournament and an annual garden tour in Rockford.

When
1997 and annually

Contact
Lisa J. Brown
Marketing Specialist
Rockford Mass Transit District
520 Mulberry St.
Rockford, IL 61101-1016
Tel: 815/961-2226
Fax: 815/961-9892

TRIP Employer Pass Program (C-10)
Kansas City Area Transportation Authority

Number of Vehicles: 243 buses

Strategy

The Transit Rider's Incentive Plan (TRIP) is the employer pass program conducted by the Kansas City Area Transportation Authority (Metro) of Kansas City, MO. TRIP allows employees to conveniently purchase monthly bus passes at work tax-free. The Metro provides a four dollar discount on each purchase. Employers are asked to match the amount and are encouraged to contribute more as an employee incentive. Incentives offered to TRIP participants include the Emergency Ride Home Program, which guarantees a free taxi ride home when an emergency arises. The Metro arranges and pays for the taxi service.

Three strategies are used to implement TRIP: advertisement, one-on-one, and media relations. For advertising, TRIP utilizes interior and exterior bus advertising. Bus riders contact the agency and distribute information about the program to their employers. The employee's initiative has a big impact on employer's interest in the program. Follow-up calls are made and a company profile is developed which includes the type of business, its location, if employees work at the same address, total number of employees, the contact name and the CEO. Soon after, an interview is conducted with the Human Resources department regarding parking and transportation situations for the particular employer.

Media relations prompt new business because firms identify with what their competitors are offering their employees. Local news articles are printed periodically to introduce the program and to thank new TRIP members for joining.

Promotional materials are produced by the Metro and are given to each TRIP outlet. Materials are customized for organizations to introduce the service to employees. Informational sessions are conducted to help employees learn how to best use the system.

Objectives

To increase ridership by attracting new businesses to TRIP and to build pass sales within current member firms.

Resources

TRIP works within an annual budget of $10,000. One agency staff member is required to maintain the program.

Implementation Time

One to four weeks, depending on the size of the employer.

Results

Employer participation in TRIP increases about 20 percent every year. Success is measured by the total number of monthly bus passes sold and the total number of member firms.

When

1991 and continuing

Contact

Cindy Baker
Director of Marketing
Kansas City Area Transportation Authority
1200 E. 18th St.
Kansas City, MO 64100
Tel: 816/346-0209
Fax: 816/346-0305

"Transit Works!" (C-11)

Number of Vehicles: 900 buses

Strategy

Until 1991, Metro Transit of Minneapolis, MN sold transit fares to employers at an 8.3 percent discount that could be passed along to employees. In that year, 450 employers participated in the program. When the agency introduced "SuperSavers," transit fares were made available to the public at a 30 percent discount. By 1996, the number of employers offering transit passes through payroll deduction had dropped to 215. As Metro Transit introduced new magnetic fare media in mid-1996, it began marketing a renewed employer-based program that offers a discount in addition to the SuperSaver discount. By state statute, only employers offering transit passes through payroll deduction could participate in the program.

Metro Transit developed a three-part campaign to attract new interest in its free employer pass program. "Transit Works!" was chosen as the program's title to tie it closely to the work commuting market and to put a positive spin on transit use. The first component was an informative brochure mailed to employers. This piece included a return card for inquiries. The second tactic was a brochure mailed to employees. This lead-generation piece contained two reply cards: one that could be sent interoffice to the employee's human resources department requesting investigation and another that could be sent to the transit agency, which would cause the forwarding of an information packet to the employee's company. Metro Transit also contacted employers previously or currently enrolled in payroll deduction to re-introduce the program and established a hotline that prospective participants could call for information.

Objectives

To reinvigorate the transit agency's employer pass program.

Resources

Metro Transit operates the program within an annual budget of $10,000.

Implementation Time

Six months

Results

By the end of 1996, Metro Transit had received letters of agreement from more than 300 employers. The agency received nearly 200 reply cards for information, both from employers and employees. Several employers, many of them smaller businesses, were interested in offering the program but declined participation, citing difficulty in establishing a payroll deduction system. These concerns were alleviated by the Minnesota state legislature, which revoked the payroll deduction requirement in 1997.

When

1996 and continuing

Contact

Kathy Laudenslager
Market Development Specialist
Metro Transit
560 Sixth Av. North
Minneapolis, MN 55411
Tel: 612/349-7531
Fax: 612/349-7675
E-mail:
Kathy.Laudenslager
@metc.state.mn.us

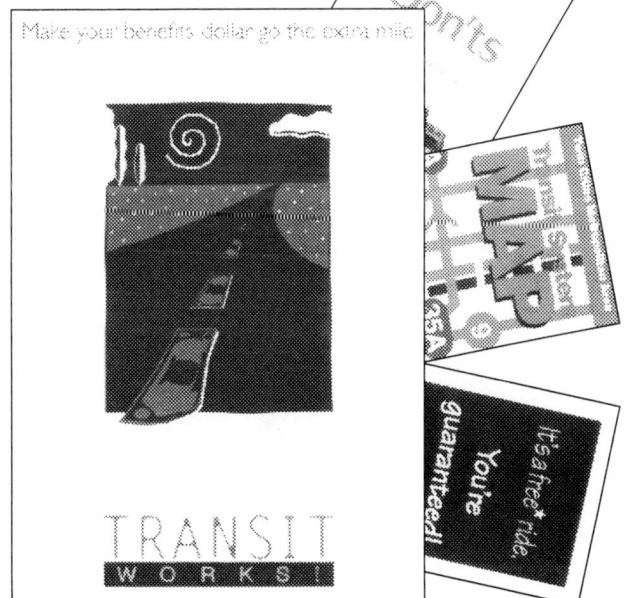

FoxTrot Promotion (C-12)

Number of Vehicles: 22 buses

Strategy

Transfort of Fort Collins, CO promoted the advantages of its new regional bus service between the cities of Fort Collins and Loveland by joining with five local merchants to provide discounts or free merchandise to holders of monthly FoxTrot passes. Transfort officials approached various businesses along the route and asked them to offer discounts in exchange for being named in all advertising for the passes, such as bus signs, flyers, newspaper and radio ads.

Objectives

To promote the advantages and conveniences of the new service to both riders and affected merchants.

To increase sales of monthly passes.

Resources

The costs of the co-op promotion covered flyers, newspaper and radio ads, and staff time.

Implementation Time

Two months

Results

Transfort believes the promotion has been successful. The agency hopes to include more merchants in the program.

When

June 1997 and continuing

Contact

Linda Gale Dowlen
Transportation Demand Supervisor
Transfort
6570 Portner Rd.
Fort Collins, CO 80525
Tel: 970/224-6191
Fax: 970/221-6239
E-mail: LDowlen
@ci.fort-collins.co.us

Rider Incentives (C-13)   *VIA Metropolitan Transit*

Number of Vehicles: 529 buses

Strategy

To attract new commuter riders and increase ride frequency of existing riders, VIA Metropolitan Transit of San Antonio, TX works with various companies to offer incentives to passengers in exchange for ads on interior bus cards. These "perks" are then provided to passengers in several ways and are promoted on radio ads and through in-bus flyers.

The procedures for each incentive varies. In all cases, VIA exchanges space in the interior of the buses for a privately provided benefit to riders. The sponsors of the "perk" provides the bus cards, which are placed in 60 to 200 buses for a limited time. VIA radio ads are tagged with the current promotion. The various incentives are given to passengers through contest drawings, random ticket giveaways, or as bonuses when buying a monthly pass. In the first few months of the program, 1,000 movie tickets, circus tickets, and discount cards for local malls were distributed.

Objectives

To create a fun atmosphere for bus passengers and create a less serious perception about riding the bus.

To increase transit ridership.

Resources

The only cost to VIA is the printing of in-bus flyers, which range from $100 to $1000, depending on the incentive.

Implementation Time

Two months

Results

VIA believes the promotion is an easy way to give added value to passengers, and does not compete for significant funding from the agency's limited budget. In giving away the tickets and speaking with passengers, the agency discovered the promotion is being very well-received. Numerous inquiries from potential additional sponsors have been received.

Adaptations

VIA has expanded the program to include riders with different types of passes.

When

September 1997 and continuing

Contact

Steve Cerna
Public Affairs Coordinator
VIA Metropolitan Transit
800 W. Myrtle
San Antonio, TX 78212
Tel: 210/362-2378
Fax: 210/362-2572

United Shoppers Service (C-14)

Number of Vehicles: 46 buses, 2 trolleys, 12 vans

Strategy

A major local grocery chain, United Supermarkets, underwrites the cost of providing public transit service directly to retirement homes, community centers, and private residences along the routes carrying seniors and physically challenged individuals to the supermarket for shopping. The service runs five days a week, twice on one day. These are routes "guaranteed" by the supermarket chain, although run open to the public. Lift-equipped full-size transit vehicles are used. The vehicles run with a "United Shopper" header and no fares are collected while the route is underway.

Objectives

To reinvest in the community by providing transportation for elderly and physically challenged citizens for supermarket trips.

Resources

The transit system approached the supermarket initially about underwriting the service from retirement homes to the supermarkets. The supermarket chain has continued to underwrite the service throughout the years. The cost to United Supermarkets is estimated to be $400 per week.

Implementation Time

The service requires minimal staff time to maintain.

Results

The transit system has judged the strategy not only to be beneficial in terms of the needed service provided to the citizens in its service area, but also the public relations and image building gained from media exposure. The service usually receives news coverage on radio, television, and print media generally at least once a year. Not only do calls requesting information about the specific service go up after news coverage, but the public transit system continues to get kudos from the general public for the service.

When

Ongoing (initiated more than 25 years ago)

Contact

Scott L. Mitchell
Marketing Director
Citibus of Lubbock
PO Box 2000
Lubbock, TX 79457
Tel: 806/767-2380, x 234
Fax: 806/767-2387
E-mail:
Smitchell@citibus.com

Weekend Shopper's Shuttle (C-15)

County Ride

Number of Vehicles: 8 buses

Strategy
County Ride, Queen Anne's County, MD public transit agency, has initiated a weekend shopper's shuttle service in the Kent Narrows area through a joint effort with 50 local businesses to encourage the creation of a tourist attraction area. The businesses in the prospective area each contribute $50-$100 per month to pay for bus service. In exchange, the businesses can place advertisements on seat backs and commercials on a tape playing in the bus that points out historic and scenic information along the route. Tee-shirts for the project, bus stop signs, and all operating costs are covered by the business contributions.

Objectives
To assist in creating a new tourist area.

Resources
The entire project is funded by local business contributions.

Implementation Time
Six to nine months

Results
The service is paid for without any taxpayer subsidy for operating and marketing costs. The project also raised the visibility and support for County Ride in the local business community.

When
1991 and continuing

Contact
Sue Leager
Director
Queen Anne's County
Dept. of Aging
104 Powell Street
Centreville, MD 21617
Tel: 410/758-0848
Fax: 410/758-4489

Image
Promotions

mage promotion builds connections between the transit agency and the community. The goal of an image promotion is to present a positive image of the transit industry in general and the agency. Unlike a promotion about a specific service or route, image promotions are often targeting the community at large. It is a good opportunity to present the benefits of the publicly subsidized transit system to the community.

Bus Naming Contest (D-1)   *People Mover*

Number of Vehicles: 42 buses

Strategy
People Mover of Anchorage, AK received 18 new buses in 1996, the first new buses acquired by the system since 1983. With their arrival, the agency decided to celebrate with a bus naming contest open to all grade school children in the city.

Permission was obtained from the school district to conduct the contest. Entry forms were sent to each class with a two-week deadline for submission. Agency staff selected the winning names and the winners were notified.

On the day of the celebration, the winning bus was taken to the school

from where its name was submitted and the student, classmates, teachers, and parents were given a ride to the ceremony. The Mayor of Anchorage made the presentations. Standing in front of the winning buses, the students were given a small plaque showing the bus name and the student's name and had their picture taken. Each winning class was given an in-classroom lunch from Pizza Hut on a date selected by the teacher.

Objectives
To involve the community in celebrating the arrival of new buses.

To increase community awareness of public transit.

Resources
The total cost of the project was $1,375. Total staff time was 152 hours. Pizza Hut donated the in-classroom lunch for each class of the winning students.

Implementation Time
Two months

Results
More than 450 names for the 18 buses were submitted, far exceeding expectations. On the January day when the buses were unveiled and students awarded, it was -18° F outside, yet more than 800 people attended the ceremony. One class wrote a song that was performed during the ceremony. People Mover intends to repeat the pro-

ject when the next round of new buses arrive beginning in December 1998.

When
1996

Contact
Robert Kniefel
Public Transportation Director
People Mover
3650-A E. Tudor Rd.
Anchorage, AK 99507-1252
Tel: 907/343-8402
Fax: 907/563-2206
E-mail:
crazybob@alaska.net

Thumbody Express-ions (D-2) — *Caro Transit Authority*

Number of Vehicles: 6 buses

Strategy

The Caro Transit Authority of Caro, MI, the governing body of the Caro Thumbody Express, publishes a quarterly newsletter called "Thumbody Express-ions." It informs the general public about public transportation, advertises community events, current information about the community, and a bit of nostalgia. The newsletter is distributed to businesses, schools, and area organizations.

Targeted groups focused on in the newsletter include: workers, schoolchildren, working parents, and local retailers. Children's organizations such as scouting and 4-H are promoted in the newsletter. Retailers receive mention and are informed of the number of passengers dropped-off at their location.

Objectives

To change the public perception of the local bus system.

To inform the public of the advantages of public transportation.

To increase support of the bus system from the local business community.

To increase ridership.

Resources

Federal and State marketing grants for rural transit agencies are used to print the newsletters. Contributions of articles and information are accepted from the local community.

Implementation Time

Eight to ten hours per issue.

Results

Once known as the "handicapped bus," the service is now accepted as a general public transit service. There is increased support from the agency within the business community. Mileage has been increased, and the passenger count is stable.

When

The newsletter is published quarterly.

Contact

Jennifer Leitzel
Manager, Caro Thumbody Express
Caro Transit Authority
317 S. State St.
Caro, MI 48723
Tel: 517/673-8488
Fax: 517/673-7310
E-mail:
CTA@centuryinter.net

Video for Speaker's Bureau (D-3) *Good Wheels, Inc.*

Number of Vehicles: 72 buses

Strategy

Good Wheels is the state recognized provider of transportation for individuals with physical challenges in Lee and Hendry counties in Florida. In its six-year history, demand for its services has doubled from 132,000 riders to 259,000 riders. Despite this growth, Good Wheels was not well known in the community. In January 1995, agency officials retained Susan Bennett Marketing and Media to help improve its image and develop an ongoing public relations and marketing plan. The public relations agency recommended a public relations strategy that would place Good Wheels in a proactive position with the overall goal of creating an image as a responsive, well-managed non-profit transportation company. One component of the plan was the development of a video to serve as the mainstay of a new speaker's bureau.

The theme of the video, "The Wheels of Independence," was used to demonstrate how Southwest Florida's physically challenged populations have attained transportation independence because of Good Wheels.

The show was scripted and put out for bids to get the most cost-effective production. A local cable company was the low bidder. Filming occurred over one week, with two days of editing.

Objectives

To position Good Wheels as the leading transportation provider for the physically challenged in Southwest Florida.

To illustrate Good Wheel's growth over its six-year existence.

To highlight the social service agencies that depend on Good Wheels for transportation of their clients.

Resources

Scriptwriting and coordination was conducted as part of the contract with the public relations firm. Production and editing of the video was $2,600. The local cable company donated $2,000 in production services to the project.

Implementation Time

Three to four months.

Results

Response to the video was immediate and gratifying. More than 30 presentations were made to civic groups and organizations in the first 6 months after production.

Upon seeing the video, community leaders were impressed with Good Wheels growth and were unaware that the agency provides so many rides to physically challenged individuals and serves so many social service organizations. A public service announcement was derived from the video, and both of them received Awards of Distinction from the Florida Public Relations Association for meeting stated objectives. Good Wheels continues to present the video at speaker's bureau engagements, trade show booths, and as part of presentations to county funding entities.

Adaptations

The video was updated once at a cost of $600. It will again be updated to address welfare-to-work issues.

When

1995 and regularly presented since then.

Contact

Deloris Sheridan
President
Good Wheels, Inc.
10075 Bavaria Rd. SE
Fort Myers, Fl 33913
Tel: 941/768-6184
Fax: 941/768-6187
E-mail: gowheel@aol.com

"On the Move" Newsletter (D-4)   *Good Wheels, Inc.*

Number of Vehicles: 72 buses

Strategy

Because of the agency's high rate of growth, Good Wheels of Fort Myers, FL initiated a newsletter to communicate with passengers, funding entities, and community leaders. The non-profit agency turned to its public relations firm to create an eye-catching newsletter that imparts the latest information about Good Wheels as well as profile riders and members of its board of directors.

The newsletter ranges from four to six pages each issue with a page size of 8-1/2" by 11". This format was selected because it is cost effective, easy to mail, and does not take up much room when read on agency vehicles. Every issue has at least one article pertaining to new routes or explanations of old ones, along with fare information. To put a more human face on Good Wheels, each issue contains profiles on employees, a social service agency served by Good Wheels, a member of the board of directors, as well as passengers. Letters from riders, announcement of coming events, and ridership statistics complete the package.

Objectives

To develop a newsletter for riders, community leaders, and funding agencies with a goal of printing 1,300 copies in quarterly issues.

To create a name and look for the newsletter that reflected the high growth of Good Wheels.

To increase ridership.

Resources

The writing and coordination of the newsletter is part of an ongoing contract with the agency's public relations firm. Printing, postage, artwork, design, and production costs are approximately $2,300 per issue.

Implementation Time

The agency considers the newsletter a continuous project.

Results

The newsletter is published three times a year with good success. Ridership continues to grow. The newsletter has been honored by the Florida Public Relations Association with a Judge's Award for cost-effectiveness.

When

The first issue was Summer 1995.

Contact

Deloris Sheridan
President
Good Wheels, Inc.
10075 Bavaria Rd. SE
Fort Myers, Fl 33913
Tel: 941/768-6184
Fax: 941/768-6187
E-mail: gowheel@aol.com

Changing Paint Schemes (D-5)   *Citrus Connection*

Number of Vehicles: 30 buses, 13 mini-buses

Strategy

Citrus Connection of Lakeland, FL decided to change the exterior paint scheme of its buses. It was envisioned that a higher level of community support will result as more people identify with the new color scheme. The original paint scheme consisted of orange, yellow, and green stripes on a white background. The new paint scheme is a row of three colors of fruit - yellow, tangerine, and orange - on buses with different background colors. The new backgrounds include yellow, green, pink, raspberry, purple, turquoise, teal, burgundy, copper, silver, and gold.

The project was implemented by hiring a graphic design artist to create the design and then having the agency's maintenance staff reproduce it in a newly constructed paint bay.

Objectives

To change the public image of the Citrus Connection.

Resources

A capital grant was used to fund construction of the paint bay at a cost of $80,000. The graphic design artist was paid $150 for the new design. Paint and labor costs for each bus is approximately $3,000.

Implementation Time

Three years

Results

The new paint scheme has been a primary source of comment and compliments.

When

1997

Contact

Steve Githens
Transit Director
Lakeland Area Mass Transit District
1212 George Jenkins Blvd.
Lakeland, FL 33815
Tel: 941/688-7433
Fax: 941/683-4132
E-mail: CitConGit@aol.com

Star Trolleys (D-6)   *Metropolitan Transit Authority*

Number of Vehicles: 144 buses, 37 vans, 13 trolleys

Strategy

Metropolitan Transit Authority (MTA) of Nashville, TN has named four of its trolleys after famous country-western musicians. The names chosen originate from a variety of sources. Public sentiment led to one trolley being named for Chet Atkins, an opinion poll chose George Jones, and the city government designated Marty Stewart for one of the vehicles. The "Dolly Trolley" is named for Dolly Parton. A ribbon-cutting ceremony is held on the designated trolley's first day of service, with the musician and local government officials in attendance. In addition to their name being on the trolley, a glass partition in the vehicle contains an etching of the musician's signature and a quote from the artist. The trolleys operate in historic districts, such as the downtown area and Music Row, and in the Music Valley Drive/Opryland area. The transit system also provides local musicians for live entertainment aboard the trolleys.

Objectives

To enhance the trolley ride for passengers and tourists.

To increase public awareness of the trolleys in the agency's system.

Resources

The costs to the MTA are considered minimal. The glass etching, lettering and decals for the trolleys, and supplies for the ribbon-cutting ceremony are the main expenses of the program. The performers work for tips plus a minimum amount of $5 an hour.

Implementation Time

One month for each trolley.

Results

The program has been very successful and has expanded its role as a promoter of tourism. Concierges at downtown hotels have been allowed to ride the trolleys for free in order to pass along information about the trolleys to visitors. Local merchants and representatives from outlying tourist attractions board the trolleys and inform passengers about Nashville and its attractions.

Adaptations

The trolleys are utilized in a co-sponsored holiday lights tour program called "Holiday Trolleys."

When

1996

Contact

Jennifer Kocak
Rideshare Coordinator
Metropolitan Transit Authority
130 Nester St.
Nashville, TN 37210
Tel: 615/862-5961
Fax: 615/862-6208

METRO Online Website (D-7)

Santa Cruz METRO

Number of Vehicles: 93 buses

Strategy

In order to enhance its level of customer service and disseminate information about the agency more effectively, Santa Cruz METRO of Santa Cruz, CA developed and created its own website. Customers now have 24-hour, 7 days a week access to information about transit services in the county.

Objectives

To provide a continuously available source of customer service information.

To position the agency as contemporary and "high tech."

Resources

The website was developed in-house. The primary source of funds was a local air district grant of $29,000. HTML preparation was provided by a volunteer. The agency used the air district funds to purchase a computer server and pay line costs for one year.

Implementation Time

This project was conducted incrementally over a lengthy period of time. A website can be constructed by a transit agency in a much shorter period.

Results

The website is at http://www.scmtd.com and is currently providing more than 2,200 user sessions per month, 45 percent of which are in off-hours. Information is regularly updated.

When

July 1995

Contact

Mark Dorfman
Assistant General Manager
231 Walnut Avenue
Santa Cruz METRO
Santa Cruz, CA 95060
Tel: 408/426-6080
Fax: 408/426-6117
E-mail:
mdorfman@scmtd.com

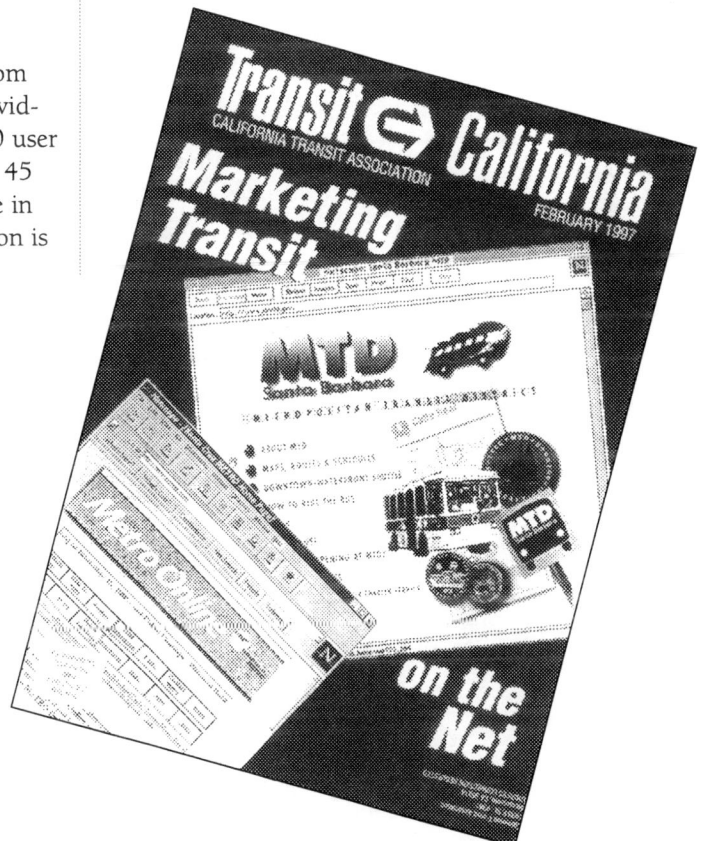

"Pledge to Our Customers" & Customer Service Tour (D-8)
MARTA

Number of Vehicles: 704 buses, 240 rail vehicles

Strategy

The Metropolitan Atlanta Rapid Transit Authority (MARTA) of Atlanta, GA has developed a new, agency-wide customer service program. Two parts of the program meet the criteria of cost-effectiveness and could be utilized by any transit agency. The first is the MARTA "Pledge to Our Customers." It is a document created and approved by MARTA's Board of Directors. It is distributed to existing and potential customers. The pledge outlines 10 basic qualities of service standards MARTA's customers should expect from the agency every day. By giving it to customers, employees are accountable for knowing the high standards set by the agency. The pledge encourages individual and team staff efforts to work towards realization of the goals. The program is supplemented by a "Report Card" that measures customer opinions on MARTA's customer service activities.

Supporting the "Pledge to Our Customers" program is a series of Customer Service Tours. This involves the MARTA's General Manager/CEO and a rotating team of employees, "The Customer Service Team," meeting and greeting riders throughout the MARTA service area and listening to their experiences riding the MARTA system.

Richard J. Simonetta, the GM/CEO of MARTA, was the focal point of the tours. Although the program was an agency-wide image campaign, the marketing department felt there had to be a focus on something or someone that could provide credibility to the program for customers. The staff considered a celebrity spokesperson, a mascot, and other tactics, but felt that having the head of the agency greeting customers throughout the system would most effectively impart a feeling of concern for riders and a commitment to improvements. A typical Customer Service Tour begins with the CEO and the Customer Service Teams at a bus, train, bus stop area, or rail station to greet customers. While speaking with and listening to customers, the team tries to communicate several points. The team informs customers about "the Pledge" and gives them brochures, distributes report cards, and distrib-

utes "I Met Rick" buttons which, if worn in the future and seen by the CEO on another tour, is good for a surprise (an inexpensive promotional item).

Objectives

To publicize the high caliber of leadership at MARTA.

To build public confidence in MARTA's performance.

To elevate MARTA's role in the future development of the region.

To put a customer-friendly face on MARTA.

To create a more relaxed, fun, and accessible environment for MARTA's agency and customers.

To tear down barriers between MARTA and its customers in order to make MARTA more accessible and concerned about customer needs.

Resources

By taking advantage of in-house production staff, MARTA was able to keep the estimated cost of the project to $15,000 over an eight-month period.

Implementation Time

Six weeks

Results

Both program components have been a great success with all objectives being reached. After the first seven tours, the Customer Service Teams had distributed more than 30,000 report cards to MARTA customers and employees. In the first five months, 3,500 cards had been returned by customers. MARTA con-tinues to receive 20 to 30 a day. Suggestions for improvements have been studied and several have been implemented.

When
January 1997 and continuing

Contact
Mitzi Rutledge
Senior Marketing Manager
MARTA
2424 Piedmont Road
Atlanta, GA 30324
Tel: 404/848-5515
Fax: 404/848-5098

Mural Beautification Project (D-9)
Toledo Area Regional Transit Authority

Number of Vehicles: 180 buses

Strategy
The Sea Gate Centre pedestrian concourse mural is a joint effort in city beautification by the City of Toledo, OH and the Toledo Regional Transit Authority (TARTA). The north wall of the concourse features the TARTA system. The mural's title, "Public Transportation Serving the Community, Yesterday, Today, and Tomorrow," exactly describes the images. At the beginning of the wall, there is a scene showing a horse and buggy. At the end of the concourse, a new-age TARTA skyway rail system is rendered. The idea for the mural originated in the Toledo Mayor's office as a desire to make the concourse more interesting. After an agreement between TARTA and the city was finalized, a call for ideas and sketches was made to artists in the community. A number of meetings led to a decision on which artist to use.

Objectives
To reflect TARTA's and the City of Toledo's commitment to the importance of aesthetics in Toledo's public spaces.

To promote the importance of public transit in the community.

Resources
Funding for the project came in part from a federal grant for the rehabilitation of downtown transit stations.

Implementation Time
One year

Results
Many praises for the mural have been received, some noting that the concourse is a much more stimulating walk than before.

When
The mural was dedicated in June 1997.

Contact
Bill Herr
Planning Director
TARTA
1127 W. Central Av.
PO Box 792
Toledo, OH 43697-0792
Tel: 419/245-5222
Fax: 419/243-8588

Internal Promotions

Internal Promotions

nternal promotions are an opportunity for a transit agency to enhance the organization from within, with the attendant result of strengthening the level of service provided to riders. The level of morale within a transit organization directly affects its level of customer service. Good internal promotions are effective morale boosters. They enhance the unity and confidence of staff and build good two-way communication between management and employees. Internal promotions can respond to both short- and long-term agency needs, all the while building credibility with the community and creating a positive public image.

Blue Jeans for Needy Families (E-1)   *VIA Metropolitan Transit*

Number of Vehicles: 529 buses

Strategy

VIA Metropolitan Transit in San Antonio, TX helps fund local community outreach projects through a program whereby employees are encouraged to wear blue jeans on Fridays at the cost of a $1 donation for the day. The money is collected by a designated employee, the "Blue Jean Rep," in each department. The employee is given a sticker that is dated and worn that day. The sticker reads "Blue Jeans for Needy Families." The Community Relations Coordinator at VIA is in charge of the project.

Objectives

To boost employee morale by allowing the wearing of jeans on Friday.

To help less fortunate families in the community.

To increase the positive image of VIA within the community.

Resources

The budget direct cost for the project was $200 for the stickers, which was paid from the donations. VIA's community relations coordinator devoted approximately 85 hours per year to the project. The "Blue Jeans Rep" job requires approximately 15 hours per year, while finance personnel estimated eight hours spent on the project each year.

Implementation Time

One week.

Results

VIA employees were able to contribute program donations to several community outreach projects. They include: The Adopt-a-Family Christmas Project, the March of Dimes fundraiser, the Santa Rosa Children's Hospital Miracle Network, and the Back-to-School project for San Antonio Metropolitan Ministry transitional families and neighborhood elementary schools. The program continues to expand, with an Easter basket project included in 1998. The organizations are very appreciative of the support and VIA employees enjoy participating very much.

Adaptations

Each fall, employees are encouraged to wear Blue Jean Fridays attire during VIA's two-week United Way campaign. Employees purchase the $1 stickers and the money collected for the casual days from Monday to Thursday are donated to the United Way along with other pledges at the end of the annual campaign.

When

Project began in the fall of 1995 and is continuing.

Contact

Gloria Boysen
Community Relations Coordinator
VIA Metropolitan Transit
800 W. Myrtle
San Antonio, TX 78212
Tel: 210/362-2370
Fax: 210/362-2572

Wellness Program (E-2)   *Utah Transit Authority*

Number of Vehicles: 534 buses, 60 vans

Strategy

A wellness program was developed by Utah Transit Authority in Salt Lake City, UT to help employees and members of their families choose healthy lifestyle alternatives. Services are free of charge to employees and spouses.
The program has grown from one room to eight centers at four different transit divisions. All of the centers are located in transit facilities, and include treadmills, weight sets, stair climbers, and aero-bics classes. The program also includes a yearly fitness evaluation with an annual bonus for positive results, personal training and nutrition analysis, smoking cessation and weight management programs with bonuses, both work and non-work injury rehabilitation, and an annual health fair.

Objectives

To promote wellness opportunities to transit agency employees in order to support and maintain healthy lifestyle choices and promote an overall sense of well being.

Resources

Funding for the wellness centers comes from two percent of any savings on annual health insurance premiums. The project has support from the general manager and the board of directors. The centers do not have a staff attendant, although a wellness director is assigned to the facilities.

Implementation Time

The original center was incorporated into the design of a new transit facility.

Results

Since 1984, the program has grown to include 960 employees out of 1,200 at the agency. Health insurance premiums for the agency have increased at only one-third to one-half the national averages since the program began.

When

1984 to the present

Contact

Raylene Thueson
Wellness Program Administrator
Utah Transit Authority
3600 South 700 West
PO Box 30810
Salt Lake City, UT 84130-0810
Tel: 801/262-5626 x2339
Fax: 801/287-4555

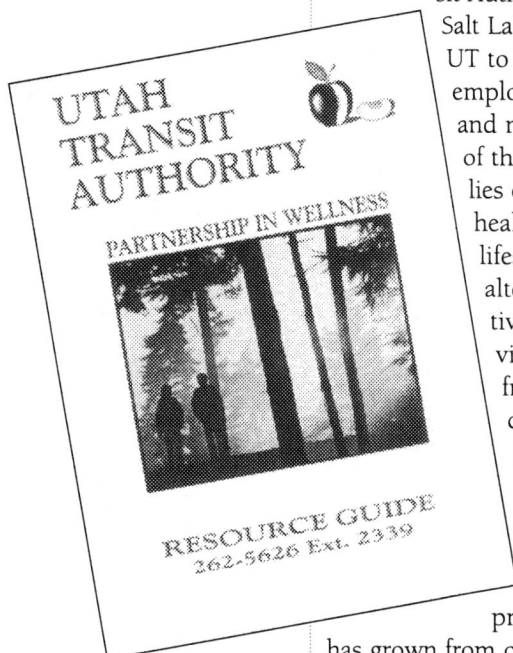

UTAH TRANSIT AUTHORITY
PARTNERSHIP IN WELLNESS
RESOURCE GUIDE
262-5626 Ext. 2339

Driver's Excellence Award (E-3)

Tangipahoa Public Transportation

Number of Vehicles: 8 buses

Strategy

In 1996, Tangipahoa Public Transportation of Amite, LA initiated an annual Driver of the Year award. In order to arrive at their decision, the managers of the transit agency used a three-part evaluation system. The first part consisted of a supervisor's appraisal of the driver's job performance. The second part was an evaluation of the driver by the Transportation Coordinator. Finally, passenger surveys were sought to gauge the driver's abilities. The driver receiving the best cumulative score is designated Tangipahoa Public Transportation Driver of the Year. The award is presented at an annual parish-wide event in conjunction with National Transportation Week. The driver's name is engraved on a plaque that is displayed at the office from which the driver works. The driver also receives a $25 cash award.

Objectives

To make drivers more aware of the importance of their roles as transit providers.

To promote confidence, credibility, and a rapport with the community.

Resources

The cost of the project is under $80, including the cash award, plaque, certificates, and frames. No significant additional staff time is required.

Implementation Time

Information is gathered from July to April.

Results

The project provides the added bonus of feedback from the rider surveys. The agency feels it receives a good indication of service satisfaction levels. This is an ongoing project.

When

1996 and continuing

Contact

Mary Dowling
Transportation Coordinator
Tangipahoa Public Transportation
106 Bay St.
Amite, LA 70422
Tel: 504/748-7486
Fax: 504/748-3199

On-Site Child Development Center (E-4)
VIA Metropolitan Transit

Number of Vehicles: 529 buses

Strategy
VIA Metropolitan Transit in San Antonio, Texas operates an on-site Child Development Center (CDC) providing quality care for the children, grandchildren, nieces, and nephews of VIA employees. The CDC provides high quality childcare at less than market rates and is available beyond the operating hours of most commercial child care facilities. The CDC was developed as a benefit to assist in recruiting and retaining highly qualified employees. VIA believes that dependable, quality childcare that is available during extended hours (Monday-Friday, 5:30 a.m. to 7:00 p.m.) reduces absenteeism and turnover among employees. The Center provides employees with a sense of security and well-being regarding the welfare of their children.

The Center accepts children from birth to seven years old. It is managed and operated by a staff of highly qualified early childhood educators who are VIA employees. The CDC can accommodate 73 children at any one time, however due to drop-in care and extended operating hours, more than 73 children can actually be enrolled. A parent advisory committee helps to shape policies and activities as needs and conditions change.

Objectives
To aid in recruiting and retaining highly qualified employees.

To reduce absenteeism and turnover among employees.

Resources
The total cost of the CDC was $400,000, of which more than $310,000 was for construction. A feasibility study was conducted on behalf of VIA by a child care consulting firm, along with site evaluation and financial proformas. To construct the CDC, VIA remodeled an existing, unused building on its property. Federal Transit Administration formula capital grant funds were used for 80 percent of the construction funds.

Employees using the center pay the majority of operating costs through weekly tuition. Operating costs are also subsidized through VIA's general operating budget. During the first year of operation, VIA's subsidy was 49 percent of direct costs. By 1996, 22 percent of operating costs were subsidized by VIA. The center also holds various fundraising activities throughout the year.

Implementation Time
Ten years from conception to implementation; three years from decision to implement to opening facility.

Results
The CDC has been continuously operating at or near capacity since it opened. It continues to receive widespread support from parents and non-parents alike. VIA believes that absenteeism and employee turnover have been reduced, productivity has increased, and the Center is a valuable tool in the recruitment of employees. Employees who do not use the Center benefit from the well being of fellow employees who are utilizing the service. The fundraising activities are morale boosters for the agency, and contribute to a positive work atmosphere.

When
The CDC opened in September 1990 and has been in continuous operation since.

Contact
Mary Helen Santillan
Child Development Center Administrator
VIA Metropolitan Transit
800 W. Myrtle
PO Box 12489
San Antonio, TX 78212
Tel: 210/362-2230
Fax: 210/362-2581

VIA for Life Health Fair (E-5)

VIA Metropolitan Transit

Number of Vehicles: 529 buses

Strategy

VIA Metropolitan Transit in San Antonio, Texas holds an annual health fair for employees, immediate family members, and retirees. The two-day fair offers direct contact with a wide variety of health professionals and service providers. The health fair is a good opportunity for employees to become more educated about health issues. Tables and booths are set up to offer information and screenings. Screening results can identify health problems before they become serious. Over $200 worth of health screenings are free to participants. Some tests, such as blood profiles, prostrate screening, mammograms, and flu shots, require the employee to pay a modest fee.

Objectives

To provide employees and their families with the opportunity to become more aware of health issues.

Resources

The health fair is conducted voluntarily by the agencies and businesses that set up booths and tables. In most cases, the exhibitors consider this part of their mission and expect to take part in such activities at their own expense. VIA spends $2,100 to pay for health screenings that are not offered free.

VIA budgets $5,000 for the fair in anticipation of maximum participation, but usually spends approximately $3,000 per year. Two VIA staff members are utilized to contact and coordinate the exhibitors. To increase participation, VIA provides door prizes, most of which are donated, and an incentive item to employees completing a Health Check card. VIA spends approximately $900 on the incentive item, which is changed each year. VIA staff volunteers are used during the two days of the fair.

Implementation Time

Two and a half to three weeks.

Results

The health fair has grown over the years with more exhibitors and more employees and family members participating each year. Ninety-five percent of the participants are active employees. VIA believes the fair contributes to lower costs for employee health care, lower absenteeism, and higher morale.

When

The Health Fair is an annual event.

Contact

Yogi Cruz
Manager of Employee Benefits
VIA Metropolitan Transit
PO Box 12489
San Antonio, TX 78212
Tel: 210/362-2000
Fax: 210/362-2571

Introduction of New Service

Introduction of New Service

T he introduction of new service is a time for celebration at a transit agency, both internally and, if the change is large enough, publicly. The agency is growing and improving to meet new customer demands and is seen as responding to the city's development. The agency is evolving with its market. There are several ways to promote new services and many purposes that can be served. A strong promotion can get a new route or service off to a vigorous start. Local businesses that are positively affected by the new service are often willing to cooperate in promoting the service. Promotions can also reward current riders and attract new ones.

Free Fun-Filled Fridays (F-1)

Metro

Number of Vehicles: 379 buses

Strategy
Metro is a non-profit, public service, operating division of the Southwest Ohio Regional Transit Authority in Cincinnati, Ohio. As part of efforts to call attention to a commuter service route between downtown and an upscale suburban community, Metro, Deerfield Township, and area merchants provided free fares, gifts, and live entertainment on Friday nights throughout the first

summer of operations. Local merchants sponsored fares and provided giveaways, while the Deerfield Township government (using its economic development staff) acquired entertainment and food. The promotion was advertised

primarily through flyers and interior ads on buses, while also utilizing posters in area businesses and publicity in targeted media, mainly local community press. The live entertainment was assigned to random buses on the route. Tran-

sit agency staff rode the bus route handing out food and gifts.

Objectives
To promote the commuter service.
To make clear the importance of transit for business and government in hopes of securing additional funding for the route.
To reward current riders and attract new ones.

Resources
There was no formal budget for this promotion. Most of the handouts were donated. The

FREE FUN FILLED FRIDAYS
brought to you by Metro, Deerfield Township and area businesses.

Adaptations
The promotion made clear the value of daily personal contact with riders for future promotions.

When
Summer 1997

Contact
David Etienne
Communications
Supervisor
Metro
1014 Vine St., Suite 2000
Cincinnati, OH 45202
Tel: 513/632-7522
Fax: 513/621-5291

Implementation Time
One month

Results
In terms of riders, the program far exceeded expectations. Rider satisfaction is high. The program created a following, developing cachet with riders.

bands used for entertainment performed at no charge. Staff time was at least three hours each Friday through the summer. Local merchants sponsored fares and provided giveaways.

Trolley Fiesta (F-2) *City of Albuquerque Transit Department*

Number of Vehicles: 122 Buses, 6 Trolleys

Strategy

In July 1996, the City of Albuquerque, New Mexico acquired four new trolley buses, raising the number in its fleet to six. The acquisition of these buses is an effort by the city to provide a unique and enticing transportation service along the main corridor of the historic downtown area along old Route 66.

To promote the new service, the transit department hosted a Trolley Fiesta. The event was planned and implemented with the cooperation of five neighborhood and merchant associations for the purpose of introducing the city's new trolley buses to the community. The Fiesta consisted of an opening ceremony, with the Mayor as the keynote speaker, and an evening event. All trolley rides were free with transit staff on each trolley giving out buttons and literature on the trolley service and Route 66.

Participating merchants along the route displayed Trolley Fiesta banners, balloons, decorated storefronts, and offered discounts to patrons. Each participating neighborhood sponsored an activity including an antique car show, live entertainment, art shows, children's shows, etc.

Trolley riders during the Fiesta were also eligible for prizes by random drawing from major sponsors and participating merchants. They were also provided with buttons in recognition of their participation that made them eligible for merchant discounts and random drawings. To facilitate ridership during the Fiesta, the transit department provided park-and-ride facilities to riders.

Objectives

To increase ridership along Route 66 with the introduction of the new trolley buses.

To increase patronage of the local merchants.

Resources

The project budget was $6,000. Volunteer time was 10 hours. Two months of transit staff time was required.

Implementation Time

Five months initially, 3-4 months subsequently.

Results

During the two-day event, more than 4,000 individuals rode the trolley buses. Success was considered so great that it has become an annual event. It is now coordinated by the neighborhood and merchant associations, with the agency only supplying the trolleys.

When

Conducted annually in May.

Contact

Marie Morra
Marketing Manager
City of Albuquerque
Transit Department
601 Yale SE
Albuquerque, NM 87106
Tel: 505/764-6183
Fax: 505/764-6146
E-mail: mmorra@cabq.gov

Magic Bus Design Contest (F-3)   *Lee Tran*

Number of Vehicles: 43 buses

Strategy

Lee Transit was awarded an $8.26 million transit corridor grant from the Florida Department of Transportation. The grant included the purchase of nine buses and their operation for four years to ease congestion on the 18-mile corridor of US

41 between North Ft. Myers, FL and San Carlos Park, FL. To bring positive attention to the troubled corridor, Lee Tran developed the idea of a design contest for each of the nine buses. The participants had to be residents of Lee County and were to use an aquatic theme, reflective of the county's tropical setting on the Gulf of Mexico. Lee Tran sought and received support for the contest from a local radio

station, 96K-Rock. Compu-Labs, a local merchant, co-sponsored the project and donated a computer to each winner. An outline of the bus was placed in the Ft. Myers newspaper and entries were sent to the Lee Tran administration office. Over 3,000 entries were submitted. The quality of the drawings was such that 100 entries could easily have been considered strong contenders. The entries represented all segments of the community, from pre-schoolers to senior citizens. The winners are able to see "their" bus frequent the busy corridor, six days a week, over the next year. The winner's name and the co-sponsor's logo were on the back panel of each of the special buses.

Objectives

To increase the use of public transportation by tourists and locals.

To relieve congestion of the busy stretch of US 41.

To promote an environmentally friendly locale through mass transit.

Resources

Approximately $87,000 was spent on the project. Two-thirds of the money spent was on the supergraphics for the buses. The bulk of the rest was spent on radio promotion of the contest, the entry

form, and commemorative T-shirts. The shirts were given out at the unveiling and read "Poetry in Ocean, Lee Tran in Motion."

Implementation Time

Six months.

Results

Over 3,000 entries were received, many of high quality. Public awareness of the transit system increased. Lee Tran felt that the design contest would serve as a foundation for an identity and a positive presence for the existing Lee Tran system. The buses are easily recognizable along their route, usually identified by the main feature of the design. Once the buses were in service, television, radio, and newspaper advertising was used, with the greater amount spent on television. The agency has also noticed that advertisers are becoming more creative in their use of Lee Tran buses.

When

1997

Contact

Joanie Glance
Marketing Director
Lee Tran
10715 E. Airport Rd.
Fort Myers, FL 33907
Tel: 941/277-5012 x2223
Fax: 941/277-5011
E-mail: glancejk
@bocc.co.lee.fl.us

Media Relations

he purpose of good media relations is to promote understanding, goodwill , and acceptance of transit by the public. Utilizing local media to promote transit events or news is a very effective method of disseminating information. Establishing solid lines of communication with local media representatives can assure more fair and accurate coverage when the agency faces a crisis or when important transit-related news occurs to ensure fair and informed media coverage.

Radio Shows (G-1) — *Citrus Connection*

Number of Vehicles: 30 buses, 13 mini-buses

Strategy
The transit director of the Citrus Connection acts as the host of two five-minute radio programs, "Community Connection" and "Lakeland Low Down." The programs run twice a week on two AM radio stations in the Lakeland, FL area. The programs are interviews with individuals who are working on community projects or for non-profit agencies. The representatives are contacted and asked if they would like to promote their organization or event. The radio shows attempt to establish Citrus Connection as a popular institution in the community.

Objectives
To work cooperatively with other community organizations.

To maintain positive attitudes about Citrus Connection and correct any negative ones.

Resources
Citrus Connection pays for the airtime for the programs; however, both radio stations promote the shows at no cost.

Implementation Time
Approximately one day per show.

Results
Many positive comments have been received by the agency. Relations between the radio stations and the agency are very good.

When
1994 and continuing.

Contact
Steve Githens
Transit Director
Lakeland Area Mass Transit District
1212 George Jenkins Blvd.
Lakeland, FL 33815
Tel: 941/688-7433
Fax: 941/683-4132
E-mail: CitConGit@aol.com

"Smile on Monday" Contest (G-2)   *Citrus Connection*

Number of Vehicles: 30 buses, 13 mini-buses

Strategy
Citrus Connection has a co-promotion with a local radio station called "Smile on Monday." The contest requires customers to register using a form in the agency's newsletter, "The Reader's Digest." The form also contains a small survey (three questions) regarding ridership and service by the Citrus Connection. Customers can give the form to any bus operator or mail it in. A random drawing is held each Monday and the winner is announced on the local AM radio station and prizes awarded.

The project is promoted by 30-second spots on the radio station and in ads in "The Reader's Digest."

Objectives
To increase the level of support for Citrus Connection and to acquire customer feedback at the same time.

Resources
The only expense for the agency is the printing of the newsletter. The participation of the radio station, which includes prizes, is covered by a trade-out arrangement.

Implementation Time
Four weeks

Results
The promotion is considered to be very success-

ful. It has created a strong relationship between the agency and the radio station.

When
The promotion is in its eighth year.

Contact
Steve Githens
Transit Director
Lakeland Area Mass Transit District
1212 George Jenkins Blvd.
Lakeland, FL 33815
Tel: 941/688-7433
Fax: 941/683-4132
E-mail:
CitConGit@aol.com

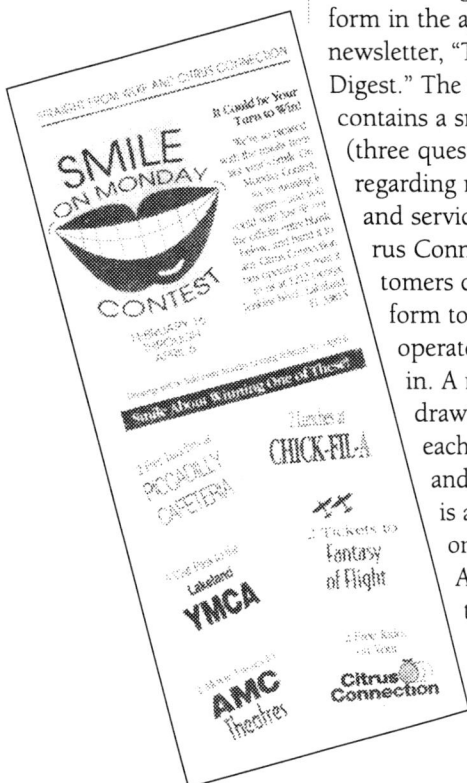

"The Big Wheel Contest" (G-3)

Wheels Transportation Services

Number of Vehicles: 18 buses

Strategy

Wheels Transportation Services of Montpelier, VT co-sponsored a listener promotion with the region's largest drive time radio program on station WDEV. Listeners who were carpooling could register with Wheels Transportation Services by contacting the agency by card, phone, or fax to request a "Big Wheel" registration form. Once the carpool is registered with the agency, it receives a gift bag and is eligible for contest and prize drawings. One registrant per carpool is allowed. An eligible carpool must have at least two riders in a car and must be driving to or from one of the 20 towns in Washington County or the towns of Williamstown, Orange, and Washington.

The gift bag contained a Wheels Transportation Services key chain, a pocket appointment book, gift certificates, a travel mug, and car wash tokens. A weekly prize of a wheel of Cabot cheese was awarded. All those registered were eligible for grand prizes of AAA memberships, a cellular phone, and an automatic car starter.

The contest received heavy radio promotion during drive time. It was also supported with press releases that included photos of weekly winners, visits to park-n-ride lots by the agency, and registration stops along major commuter routes.

Objectives

To promote the advantages of carpooling.

To enhance the agency's profile in the area.

To increase rideshare registration thereby increasing Wheels Transportation Service's state funding.

Resources

The cost of the radio ads was $960. The agency provided the gift bags, while the radio station procured the prizes. Staff time totaled 10 hours.

Implementation Time

One month

Results

The agency considers the outcome of the promotion to be mixed. While the agency received tremendous exposure, the response to the project was modest. Wheels Transportation Services received 36 new rideshare registrations, less than expected, but enough to justify the attempt.

Adaptations

In 1998, the agency conducted a similar promotion entitled "Calling All Cars." It was a more focused campaign, targeting citizens that were already carpooling regularly but not yet in the agency's database. The agency increased promotion of the campaign at worksites in the area.

When

January-February 1997

Contact

Ron Wild
Marketing Manager
Wheels Transportation Services
RR #2, Box 5650
Montpelier, VT 05602-9428
Tel: 802/223-2882
Fax: 802/223-0771
E-mail: rwild@ridewheels.org

Media Trade-outs (G-4)

County of Rockland Department Transportation

Number of Vehicles: 50 buses

Strategy

The County of Rockland Department of Transportation in Pomona, NY acquires both free and subsidized advertising for its pre-paid bus tickets on local radio and newspapers in exchange for exterior ad spaces on its buses. The agency also trades bus advertising with stores and banks who are ticket sales outlets, in lieu of commissions. Interior advertising cards are also bartered for advertisements in weekly and monthly publications.

Objectives

To extensively advertise pre-paid bus tickets and other services at little or no cost to the agency.

Resources

Advertisers pay for the production costs for sign printing and installation.

Implementation Time

Four to six weeks

Results

The agency has measured an increase in ticket sales and ridership after each promotion campaign. It estimates it has traded $70,000 to $90,000 worth of advertising since the program began.

When

May 1996 and continuing

Contact

Michael Prendergast
Marketing Coordinator
County of Rockland
Department of Transportation
The Dr. Robert L. Yeager
Health Center
50 Sanatorium Rd.
PO Box 350
Pomona, NY 10970
Tel: 914/364-2085
Fax: 914/364-2074

Media Bus Drivers (G-5)   *SporTran*

Number of Vehicles: 46 buses, 8 paratransit vans

Strategy

In order to encourage productive relationships with the local media and garner coverage of the arrival of new buses, SporTran of Shreveport, LA allows members of the media to drive buses along closed routes under controlled situations. The members of the media include local television news reporters, radio personalities, and newspaper reporters. Supervised by trainer-instructors, the media representatives handle passengers, operate lifts, use fare boxes, etc. The agency's invitations to participate include a release form. The members of the media are taught some basic rules of passenger handling with agency employees acting as customers.

Objectives

To promote the arrival of new buses.

To establish productive relations with the local media.

Resources

No direct cost was attributed to the promotion.

Implementation Time

One to two weeks

Results

The agency received coverage for days. The media personnel enjoy the experience and gain respect for the job performed by the transit agency.

Adaptations

This type of promotion can be used for almost anything: safety awareness, lift installation, driver awards, ADA issues.

When

The program has been conducted over the last 10 years when SporTran has received new vehicles.

Contact

Eugene R. Eddy
Resident Manager
SporTran
PO Box 7314
Shreveport, LA 71137
Tel: 318/673-7400
Fax: 318/673-7424
E-mail: Genee@softdisk.com

Problem-Solving Projects

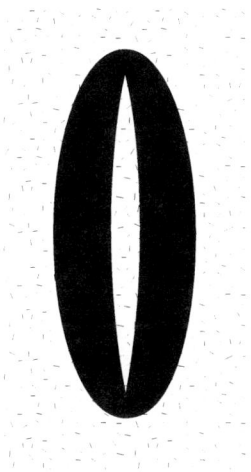

O n occasion, problems arise that a transit agency may address through marketing efforts. The problem may be one of perception by the public or internally by employees. There could also be situations that inhibit an agency from performing at its maximum level of service. A transit agency that is seen taking concrete steps to address problems as they occur, and letting people know it, retains the confidence of passengers, employees, local government, and the general public.

Bus Line Promotion Postcard (H-1) — *Central Ohio Transit Authority*

Number of Vehicles: 300 buses

Strategy
The most popular route of the Central Ohio Transit Authority (COTA) in Columbus, OH is the 2-North High Street line, that serves Ohio State University (OSU). A discreet, direct mail postcard campaign was used to inform residents living within one-quarter mile of North High Street about the high frequency of service available to them. The postcard encouraged people to take COTA to the Ohio State football games. The promotion was timed to address the problem of traffic congestion and high parking demands on game days.

Objectives
To increase ridership on the bus line and help alleviate parking and traffic congestion on the OSU campus during home football games.

To increase ridership during peak periods and raise awareness of the frequency of service.

Resources
The budget for the promotion was $21,500. More than half of the cost was for additional service. Other expenses were the design and printing costs of the postcard and postage.

Implementation Time
One to two months

Results
The bus line carried 17,357 additional riders during the six football game days, 45.7 percent more passengers than normal. Additional revenue for the six days totaled $17,078, a 76.9 percent increase over the normal level.

Adaptations
COTA uses postcards to promote new services and to inform new residents about the agency.

When
1996 and repeated annually.

Contact
Sheri Mosher
Marketing Coordinator
Central Ohio Transit Authority
1600 McKinley Av.
Columbus, Ohio 43222
Tel: 614/275-5888
Fax: 614/275-5933

Promotion of Teleride (H-2)

Number of Vehicles: 585 buses, 85 light rail vehicles

Strategy

Based on feedback from the 1995 Transit Information Use Survey, Calgary

Transit's marketing committee identified the need to promote its Teleride system and the system's new features. In 1987, Calgary Transit introduced Teleride, a computerized telephone information system that allows customers to determine when the next two or three buses will arrive at their designated stop. The system received approximately six million calls in 1994. In July 1995, new menu features were added to the system. Customers can now

call to determine future bus times and general transit information. These new features provide an opportunity to promote Teleride to customers.

The campaign ran for two months. Key messages for the campaign included the convenience of using Teleride for trip planning and the contributions transit makes to mobility for Calgarians. The primary target market for the campaign was occasional transit users. The secondary target market was potential customers.

Objectives

To provide information promoting the new features of the Teleride system.

To introduce non-users to Teleride and its features.

Resources

The cost of the promotion was $76,505. The costs included production, including the vinyl wrapping of the bus, advertising space, creative services, radio spots, and newspaper ads.

Implementation Time

Four months

Results

Customer calls to Teleride increased approximately 26 percent. A telephone survey was conducted in the last week of the campaign to measure citizen awareness of the campaign. The survey found that the promotion was successful in reaching approximately 66 percent of Calgarians. The messages were seen or heard by both Calgary Transit customers and non-customers alike, including users and non-users of Teleride.

When

October to December 1995

Contact

Rita Erven
Transportation Department
The City of Calgary
PO Box 2100, Station M
Calgary, AB T2P 2M5
Tel: 403/277-9711
Fax: 403-230-1155
E-mail: rerven @gov.calgary.ab.ca

Customer Behavior Program (H-3)   *Calgary Transit*

Number of Vehicles: 585 buses, 85 light rail vehicles

Strategy
Calgary Transit in Calgary, Alberta, Canada held discussions with its customers and employees concerning passenger etiquette on its buses. In the course of these talks, concerns were expressed about passenger behavior. These concerns included loud noise, inappropriate or offensive language, and rowdy behavior. This sort of behavior was making other passengers feel uncomfortable and, in some cases, threatened. Through research, Calgary Transit learned that some of the offenders did not realize that there are certain behaviors that are not acceptable and that these behaviors can result in fines.

The agency, with the assistance of its Youth Advisory Panel (youth aged 14-19 representing a cross-section of schools), developed a series of four messages to educate customers about the consequences of unacceptable behavior such as vandalism, graffiti, swearing, rowdy behavior, excessive noise, and eating on vehicles. The consequences of unacceptable behavior were phrased in terms that customers could relate to, such as the cost of a pair of jeans, losing $50, having to walk, or ride with their parents. An approach that supported the message in a firm but friendly tone on interior bus cards was developed. One month after the four initial interior cards, the agency followed up with a single card that thanked customers for good behavior and reinforced the original messages.

Objectives
To make customers aware of the consequences for offensive behavior.

To promote good behavior so that a ride on Calgary Transit is a positive experience for everyone.

Resources
The cost of the project was $8,700, including agency fees and production costs.

Implementation Time
Three months

Results
Based on direct feedback from customers, media coverage, and Calgary Transit employees, the campaign was considered a success. A follow-up customer satisfaction survey found a decline in the number of customers who said they felt uncomfortable because of other customer's negative behavior. Overall, the campaign was considered effective in changing customer behavior and creating a more positive customer environment.

When
1996

Contact
Rita Erven
Transportation Department
The City of Calgary
PO Box 2100, Station M
Calgary, AB T2P 2M5
Tel: 403/277-9711
Fax: 403/230-1155
E-mail: rerven @gov.calgary.ab.ca

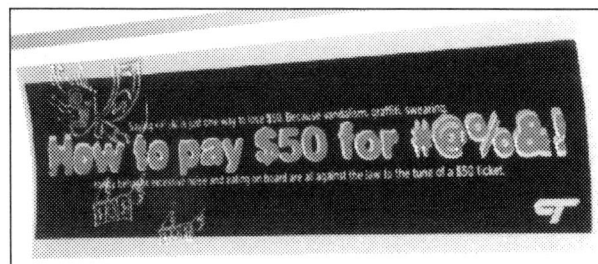

Bus Stop Blitz (H-4)

Number of Vehicles: 968 buses and vans, 53 light rail vehicles

Strategy

Dallas Area Rapid Transit (DART) in Dallas, TX uses a direct customer outreach program to smooth the introduction of service changes. Transit staff are positioned at key bus stops that provide the greatest impact and reach affected customers. The staff members are at stops during peak periods the week before the change of service and the day of the change. DART uses planners and bilingual staff at the bus stops.

Objectives

To directly communicate critical information about service changes to customers.

Resources

Twenty personnel for six hours per day for four to five days are required.

Implementation Time

One month

Results

The program is conducted for every service change by DART. The agency feels the program works especially well in areas with lower literacy rates.

When

June 1996 and continuing

Contact

Matt Raymond
Assistant Vice President, Marketing and Management
Dallas Area Rapid Transit
PO Box 660163
Dallas, TX 75266
Tel: 214/749-2801
Fax: 214/749-3668
E-mail: mraymond@dart.org

Park-N-Ride Campaign (H-5)
Regional Transportation District

Number of Vehicles: 868 buses, 17 light rail vehicles

Strategy
The Regional Transportation District (RTD) in Denver, CO provides customers with 59 Park-n-Ride locations in metropolitan Denver. The agency discovered that 37 of the Park-n-Ride lots were underutilized with utilization falling below 75 percent. The primary reason for the underutilization was that people were unaware of the lots and the services associated with them. The goal of RTD's Park-n-Ride campaign was to inform the public about lot locations in their neighborhoods. The major thrust of the campaign took place during Denver's busy summer months to capitalize on increased traffic and parking costs.

The promotion used direct mail to target "choice commuters" within a three to five mile radius of each underutilized lot. The mailing was customized for each service area and took a "neighborhood" approach to instill an ownership and community sentiment for the Park-n-Ride lot. It enticed recipients to try transit through an offer of five free rides and the free parking at the lots.

In support of the direct mail campaign, two 60-second radio ads were developed and aired just prior to and immediately after the direct mail was received in homes. The ads were broadcast during drive time hours for maximum effect. The agency also employed print ads to reinforce the overall message.

Objectives
To increase use of underutilized RTD Park-n-Rides and the routes serving them.

To increase use and awareness of all RTD Park-n-Rides and the routes that serve them.

To promote the benefits of public transportation to potential customers.

To increase public awareness of the positive effect of transit on the quality of life in Colorado.

Resources
The media and broadcast production budget for this campaign was approximately $141,000. Internal graphic designers and out source designers were used.

Implementation Time
Three to four weeks

Results
The direct mail campaign netted an 11 percent response rate to the free ride coupons. There was an increase in the use of the targeted lots and increased ridership on the express and regional routes serving them. Inquiries concerning Park-n-Ride services increased at RTD's telephone information center during the campaign.

Adaptations
The agency utilizes a similar campaign to promote awareness of its SkyRide service to Denver International Airport.

When
November 1996 and continuing

Contact
Lori Schneider
Marketing Specialist
Regional Transportation District
1600 Blake St., BLK 21
Denver, CO 80202-1399
Tel: 303/299-2023
Fax: 303/299-2008

Communibus 316 Relaunch (H-6)

Number of Vehicles: 794 buses

Strategy

A community-based campaign helped secure the future of Communibus Route 316 operated by the Ottawa-Carleton Regional Transit Commission (OC Transpo) in Ottawa, Ontario. The line operates in the downtown core on a fixed route and timetable with fully accessible 20-passenger buses and specially trained operators. It was to be cancelled in June 1996 after two years of operation due to extremely low ridership. With the assistance of regional councillors serving the area, the "Friends of Communibus," a grassroots group of users, worked with OC Transpo to redesign the route and to promote it to current and potential riders.

Working door-to-door and at bus stops, the group spread the message, "Use it or lose it." The advantages of the line were stressed, including an easily remembered hourly schedule, convenient stops at many downtown destinations popular with seniors and persons with disabilities, full accessibility, and friendly operators.

To support the volunteer promotional efforts, OC Transpo produced a flyer describing Route 316 and containing rebate coupons from retailers along the route, T-shirt and promotional buttons, an ad in the local daily newspapers, and an official launch party with refreshments served at a supermarket and a café at opposite ends of the route. Rides were free on Route 316 for its relaunch day.

Objectives

To retain service along Communibus Route 316.

To increase ridership along the route.

Resources

The total cost of the relaunch was $5,400, of which $3,000 was defrayed by advertisers in the flyer distributed by the agency.

Implementation Time

Six months

Results

The route was saved. Ridership almost doubled along the route by the end of 1996.

When

June and July 1996

Contact

Oxana Sawka
Director
Customer and Community Relations
OC Transpo
1500 St. Laurent Blvd.
Ottawa, Canada K1G 0Z8
Tel: 613/741-6440
Fax: 613/741-7359

Mural Art in Transit (H-7)　　　　　*Santa Fe Trails*

Number of Vehicles: 31 buses

Strategy

The City of Santa Fe Transit Department (Santa Fe Trails) in Santa Fe, NM was experiencing vandalism and graffiti on its vehicles. To address the issue, Santa Fe Trails joined forces with the Santa Fe Teen Arts Center: Warehouse 21 to create murals on the interior ceilings of five transit buses. The bus murals became a part of the Arts Center's 1997 Community Youth Mural Program. One professional artist and seven local teen artists created and executed five unique bus designs: Tree Bus, Olympic Bus, Heritage Bus, Stair Bus, and People Bus. During the design phase, the artists sketched designs and interviewed bus passengers to better understand the project. This pilot project displaying art created by young people is intended to reduce vandalism by garnering peer respect for the art.

Objectives

To reduce the $40,000 per year costs created by vandalism and graffiti on system buses.

To provide public art for the enjoyment of bus riders in keeping with the spirit of the arts community of Santa Fe.

Resources

The total budget was $18,500.

Implementation Time

Two months

Results

The community took an interest in this project at every stage, including the selection of the young artists, "work in progress" activities, public exhibition of the finished works, and ongoing use of the buses in transit service. Local media covered the event as a news story, resulting in positive coverage of the transit system. To date, the buses with interior ceiling murals remain free of vandalism and graffiti.

When

1997

Contact

Ana Gallegos y Reinhardt
Executive Director
Warehouse 21
1614 Paseo de Peralta
Santa Fe, NM 87501

Tel: 505/989-4423
Fax: 505/989-1583
E-mail:
agr1614sf
@webtv.net

"From the Driver's Seat" (H-8) *Sun Tran*

Number of Vehicles: 203 buses

Strategy

In order to communicate and educate passengers about expected behavior while using the bus, Sun Tran of Tucson, AZ initiated a passenger communication campaign called "From the Driver's Seat." The campaign uses a bilingual brochure and a series of 11 supporting bus posters with short messages from the bus drivers to the passengers. Sun Tran bus operators asked for a program that helps educate the public about certain aspects of riding the bus. The messages help make the operator's job easier by using inside bus cards to communicate simple messages about expected behavior, etiquette, and policies. Sun Tran believes the campaign reinforces the idea that the company and the operators want the passengers to have a safe, convenient, and comfortable bus ride.

Objectives

To communicate and educate passengers about expected bus riding behavior.

To make the driver's job easier by posting the messages.

To put a human face on the bus driver by using photos of actual operators and humor to communicate the messages.

To influence rider behavior in a positive way.

Resources

Sun Tran drafted a creative work plan for the campaign that was given to an advertising agency, for further creative design and photography. Approximately 32 project management hours were required. Printing costs were $1,000 for 5,000 brochures and $5,300 for 300 copies each of the eleven different bus posters.

Implementation Time

One year

Results

The campaign received a first-place award and a Judge's Choice Award at Tucson's annual American Advertising Awards. Passenger awareness of expected behavior on Sun Tran buses increased and the morale of bus operators was enhanced by participating in the program.

When

1997 and continuing

Contact

Sally Thompson Valenzuela
Customer Relations Manager
Sun Tran
4220 S. Park Av.,
Bldg. 10
Tucson, AZ 85714
Tel: 520/623-4301
Fax: 520/791-2285

For a fun & safe ride, know the rules of the road.

Sun Tran

Promoting Transit

Promoting Transit

Promoting transit as a viable option in the mix of transportation alternatives is essential to the success of a transit agency. Citizens in the agency's service area may be unaware of the convenience of using transit for their daily activities. Creative marketing campaigns will enhance the perception of transit as an effective alternative form of transportation and perhaps lead to new riders who may be unaware of its availability or advantages of using it. Promoting the transit service as a worthwhile public service is also helpful in attracting and maintaining support from non-users who may support that service with their local, state, and/or federal taxes.

Transportation Fair Booth (I-1) — *Auburn Transit*

Number of Vehicles: 3 buses

Strategy
In order to publicize the service provided by the agency, Auburn Transit of Auburn, CA utilized an exhibit booth at a local transportation fair. During the event, the agency gave away magnets, key chains, and route schedules.

Objectives
To inform local residents that the transit system was user friendly and served most locations within the city.

Resources
The cost of the exhibit was approximately $600. Two agency personnel staffed the booth for eight hours each during the fair.

Implementation Time
One month

Results
The agency received positive feedback regarding its efforts.

When
May 1997

Contact
Todd Strojny
Transit Manager
Auburn Transit
1225 Lincoln Way
Auburn, CA 95603
Tel: 530/823-4211
Fax: 530/885-5508

Cable Television Advertising (I-2)
Ann Arbor Transportation Authority

Number of Vehicles: 74 buses

Strategy
The Ann Arbor, MI community is served by the television stations of Detroit. As a result, there are no local stations on which to advertise. To overcome this situation, Ann Arbor Transportation Authority uses local cable television advertising. The agency runs 30-second ads on a variety of cable channels, including CNN, VH-1, ESPN, Lifetime, and the Family Channel. During college football and basketball season, the agency runs ads during all University of Michigan games televised.

The agency decides on a topic for the spot and then confers with its contracted advertising agency. A story board, script, and filming schedule are then developed. One ad was done completely in animation.

Objectives
To educate the populace served by the agency regarding the services it provides.

To increase the public visibility of the agency.

To increase ridership.

Resources
The average cost for each ad is $13,000. Staff time required for each ad is approximately 56 hours.

Implementation Time
Two weeks per advertisement.

Results
The agency measures the success of its ads through its yearly onboard survey and a general phone survey conducted by the University of Michigan every other year. Since the use of cable advertising began, the surveys have shown that both riders and non-riders recall seeing the ads more frequently than any other type of advertising conducted. In general, more than 80 percent of riders can recall the commercials, while more than 60 percent of the general public recalls the ads. Recall includes the basic theme of the ads and the message contained in them.

When
The agency began using cable television advertising in 1991. On average, one to two spots are produced every eighteen months.

Contact
Liz Nowland-Margolis
Manager of Community Relations
Ann Arbor Transportation Authority
2700 S. Industrial Hwy.
Ann Arbor, MI 48104
Tel: 734/677-3901
Fax: 734/973-6338
E-mail: liznm@theride.org

State Capitol Public Transit Display (I-3)

Nebraska Association of Transportation Providers

Strategy

The Nebraska Association of Transportation Providers increases public awareness of the services provided by the state's 59 transit systems through a 1-week display at the state capitol in Lincoln. The Association gathers photographs, passenger testimonies, and histories from the systems. This information is placed in three-ring binders and incorporated into a table display. Maps are created that display each of the systems. An Association history of activities is made available. The display is staffed for five hours each day. On one day during the week of the display, fact sheets accompanied by bus-shaped cookies are distributed to the offices of the Governor, Lieutenant Governor, and all of the state senators.

Objectives

To increase public awareness of transit agencies within the state.

Resources

Staff volunteers created the display and operated the booth. The Association paid for the materials and the cookies. The total cost to the association was less than $150.

Implementation Time

Eight weeks

Results

Reaction to the display was very positive. Citizens became more aware of the activities of other transit systems in the state. Several letters from state senators were received praising the effort.

Adaptations

The display can be used for many other events, such as fairs, trade shows, conferences, regional driver training programs, bus rodeos, etc. The Association plans to repeat the project in coordination with the Governor's proclamation for "Transit Week in Nebraska."

When

1997

Contact

Marlene Gakle
Executive Director
Nebraska Association of
Transportation Providers
1810 Sara Road
Beatrice, NE 68310
Tel: 402/223-2460
Fax: 402/223-2460
E-mail:
NE64557@navix.net

Advertising and Brochure Promoting Ridership (I-4)
Bladen Area Rural Transportation System

Number of Vehicles: 12 vans, 3 buses, 1 auto

Strategy

Bladen Area Rural Transportation System (BARTS) of Elizabethtown, NC, promoted its services through a brochure and newspaper advertisement. The campaign attempted to increase public awareness that transit services are available to all citizens in the county, not just the elderly or low-income residents. The brochure was created in-house and is distributed by drivers. The newspaper ad runs periodically.

Objectives

To increase public awareness of the full range of services offered by the transit agency.

Resources

The total marketing budget for BARTS in fiscal year 1997 was $400. All work on the promotion was done in-house. The brochures were ordered in volume through a catalog supplier.

Implementation Time

Two weeks

Results

Trips increased so dramatically that the agency was able to hire another driver. The agency's mailing list expanded and the number of callers to the system increased.

When

1997

Contact

Kent Porter
Transportation Coordinator
Bladen County Division on Aging & BARTS Transportation
P.O. Box 520
Elizabethtown, NC 28337
Tel: 910/862-6930
Fax: 910/862-6913
E-mail: doa@bladenco.org

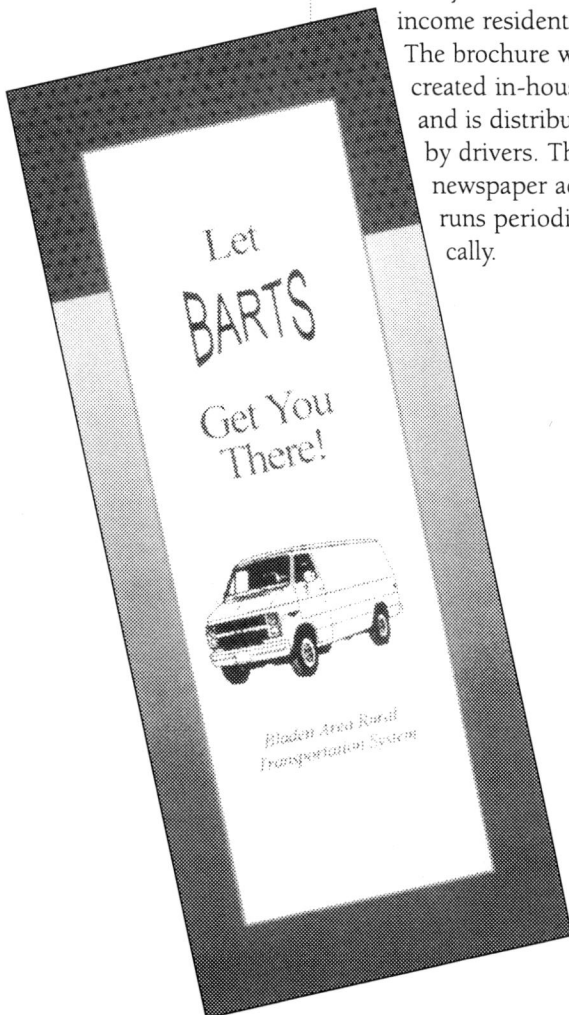

"I Have Connections" (I-5)

Kalamazoo Metro Transit System

Number of Vehicles: 40 buses

Strategy

As part of the agency's 30th anniversary celebration, the Kalamazoo Metro Transit System (Metro Transit) of Kalamazoo, MI, introduced the "I Have Connections" campaign. Passengers paying a fare were given a button. Once a month thereafter, an "I Have Connections" customer appreciation day was designated. Any passenger wearing the button was allowed to ride for free. During the customer appreciation day, new buttons were distributed to paying passengers. Customers would wear their button throughout the day, identifying themselves as Metro Transit riders.

Objectives

To reward the support of current riders of Metro Transit.

To promote the transit system's 30-year anniversary.

To increase awareness among employers and businesses that their employees and customers use Metro Transit.

Resources

The campaign theme was developed by agency staff through market research, with suggestions from the Transit Authority Board. Printed materials were developed by a local marketing firm under contract with the transit agency. There was no formal budget for the campaign. Required staff time was between 40 and 80 hours.

Implementation Time

Three months

Results

The buttons were seen on passengers throughout the Metro Transit service area during the campaign.

Adaptations

In keeping with the spirit of the campaign, Metro Transit operators requested a second button — "I'm Your Connection" — which was printed and distributed to them.

When

February to December 1997

Contact

Carmine Lewis
Administration Supervisor
Kalamazoo Metro Transit System
530 N. Rose St.
Kalamazoo, MI 49007-3638
Tel: 616/337-8408
Fax: 616/337-8211
E-mail: lewisc @ci.kalamazoo.mi.us

"Found Time" Contest (I-6)   *Transit Authority of River City*

Number of Vehicles: 277 buses

Strategy

The "Found Time" Contest conducted by the Transit Authority of River City (TARC) of Louisville, KY, used the concept that riding the bus is not a waste of time, but rather "found time." The campaign stressed that riding the bus is quality time which can be spent reading, studying, interacting with fellow passengers, simply relaxing, etc. It is time better spent than passing time, sitting in traffic, in an automobile.

TARC offered riders the chance to tell how they creatively spend time on the bus. The best entries, based on a practical yet unusual use of time, received prizes. A local supermarket chain furnished prizes of five- and three-minute shopping sprees for the first and second place winners. The third place winner received a $100 gift certificate.

The promotion was designed to leverage advertising dollars to garner free media exposure. The contest format accomplished this by using an essay entry form to attract print media and the shopping spree prizes to attract local television news coverage.

Objectives

To emphasize the concept that time spent on transit is quality time.

Resources

The total cost to the agency for the contest was approximately $30,000. Forty hours of staff time was required. Partnerships were formed with a local supermarket chain and a local newspaper to bring the campaign with the agency's limited budget for this promotion. The local newspaper co-sponsored the campaign, offering substantially discounted advertising rates in return for insertion of its logo on all printed materials and mention in radio spots. The supermarket chain sponsored the prizes.

Implementation Time

Four months

Results

The agency considered the project to be very successful. Over 1,200 entries were received. The entries were judged by a diverse group of transit and non-transit officials. The first-place winner was a member of a "club" of commuters who had been riding together for more than seven years. In her essay, she explained how the club used their commute on the bus for activities such as celebrating birthdays, baby showers, planning an annual holiday party, and other occasions.

When

October to December 1996

Contact

Perry Jacobs
Director of Marketing
Transit Authority of River City
1000 West Broadway
Louisville, KY 40203
Tel: 502/561-5118
Fax: 502/561-5253
E-mail: tarc@aye.net

Annual Meeting (I-7) *Rides Mass Transit District*

Number of Vehicles: 45 vans, 3 buses

Strategy
Rides Mass Transit District of Rosiclare, IL, has instituted an annual meeting. Along with agency staff and their spouses, the agency invites state legislators and officials from the Illinois Department of Transportation, local transportation issues and the Rides Mass Transit District. The event garners substantial local media coverage.

Objectives
To bring recognition to the services provided by Rides Mass Transit District.

To promote rural transit throughout the State of Illinois.

most effective ways to promote awareness of rural transportation in the area. It allows state and federal officials the opportunity to see the transit agency in operation and note the appreciation of the system by passengers and staff.

When
The annual meeting is held in October.

Contact
Betty S. Green
Executive Director
Rides Mass Transit District
PO Box 190
Rosiclare, IL 62982
Tel: 618/285-3342
Fax: 618/285-3340

government, and social service agencies. A keynote speaker is recruited from officials at the federal, state, or local level who have been supportive of rural transportation issues. The agency presents a "Friend of Transit" award to a person who has had the greatest impact on rural transportation in the previous year. Driver awards are also presented. The agency takes advantage of the opportunity at the meeting to inform those in attendance about rural

Resources
The costs of the annual meeting are covered through the annual budget. Renting of the room, catering, invitations, programs, and decorations cost approximately $2,600. The required staff time is approximately 25 hours.

Implementation Time
Four to six weeks

Results
According to the agency, the annual meeting has proven to be one of the

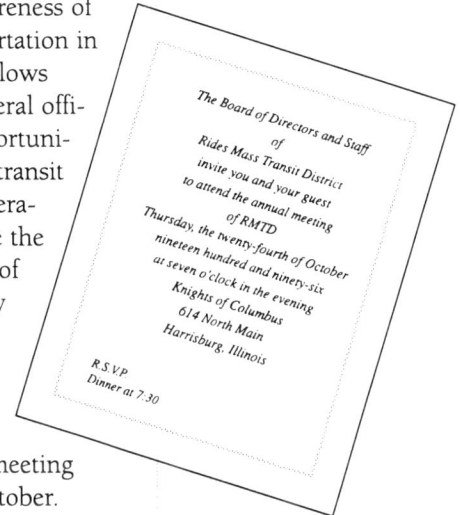

"Step Off the Gas, Step On the Bus" (I-8)   *Pierce Transit*

Number of Vehicles: 193 buses, 226 vans

Strategy

Taking advantage of a surge in gasoline prices in May 1996, Pierce Transit of Tacoma, WA, initiated a promotion allowing passengers a free ride on any agency bus route simply by showing a gasoline receipt. The one-day promotion was entitled "Step Off the Gas, Step On the Bus." The idea behind the campaign was to create an association between the high cost of driving alone and the economical alternative of transit.

Pierce Transit used an aggressive media relations campaign that generated articles in two major Seattle-Tacoma daily newspapers, live interviews on two radio stations, and coverage on local television news. The agency also ran radio ads on a Tacoma all-news station in advance of the event, and purchased underwriting announcements on the local National Public Radio station.

Objectives

To increase ridership on Pierce Transit buses.

To promote transit as an economical alternative to the automobile.

Resources

The total cost of the promotion was approximately $1,400, with most of the funds spent on underwriting the public radio ads. Required staff time was 10 hours.

Implementation Time

Five days

Results

On the day of the promotion, 875 passengers presented gasoline receipts in lieu of fares to ride agency buses, a 2.5 percent increase over average daily ridership on the system. A feature article on the promotion appeared in Passenger Transport.

When

May 1996

Contact

Jean Jackman
Public Information Officer
Pierce Transit
P.O. Box 99070
Tacoma, WA 98499-0070
Tel: 253/581-8034
Fax: 253/581-8075
E-mail: jackmanj @piercetransit.org

"Those Lextran Drivers" (I-9)

Lextran

Number of Vehicles: 47 buses

Strategy

As part of an overall strategy to reinvigorate local perception of the Transit Authority of Lexington, KY (Lextran), the agency developed a campaign utilizing a 30-second television commercial featuring Lextran vehicle operators. The commercial consisted of operators singing a catchy jingle which emphasized the increased frequency/scheduling and ease of use of the transit system. The agency believed that a combination of music and humor would have the broadest appeal in the local market. Radio ads using the same technique were also created.

Objectives

To enhance perception of the importance of Lextran to the community.

To promote Lextran's increased frequency/scheduling and ease of use.

To increase ridership.

Resources

The cost of the promotion was $30,000. The agency was able to procure $28,000 worth of free advertising from local media through a dollar-to-dollar match agreement. Lextran was also able to negotiate a spot-for-spot agreement.

Results

The agency considered the promotion to be highly successful. The commercial proved to be memorable to the public and adaptable for the campaign's target marketing needs.

When

September and October 1996

Contact

Jenny Williams
Director of Marketing and Sales
Lextran
109 W. Loudon Av.
Lexington, KY 40508
Tel: 606/255-0804
Fax: 606/233-9446
E-mail:
willjenny@aol.com

Ríder Inducements

Ríder Inducements

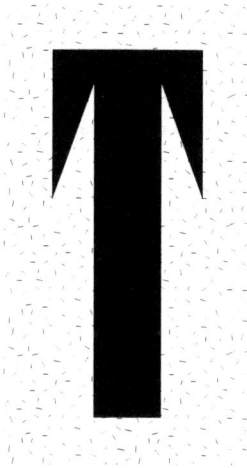

Transit agencies seek the most effective ways to induce individuals to become regular users of a transit system. Just as important as new riders, the agency must find ways to reward its regular customers. Adding value to the transit experience of core ridership helps to maintain a loyal customer base.

Free Ride Wednesdays and Saturdays (J-1)

Ben Franklin Transit

Number of Vehicles: 206 buses

Strategy
To boost ridership, Ben Franklin Transit of Richland, WA, offered unlimited free rides on its fixed-route system on Wednesdays and Saturdays. The promotion was advertised with a two-month radio and newspaper campaign. Flyers were placed in buses and posters were displayed at transit centers. The radio ads were broadcast five times a day on seven different stations over a three-month period. The newspaper ads were printed several times a week over the same period.

Objectives
To build ridership on all routes and carry it over to the entire week.

Resources
Ads for the campaign were created in-house by the marketing staff.

Implementation Time
One day per ad

Results
Ridership on the agency's fixed routes increased 14.6 percent in 1995, 11.4 percent in 1996, and 9 percent in 1997. Much of the increase is attributable to the promotion.

When
1995 and continuing

Contact
Gary Wolcott
Marketing Supervisor
Ben Franklin Transit
1000 Columbia Dr. SE
Richland, WA 99352
Tel: 509/735-4131
Fax: 509/735-1800

TransPlan Employer Information Packet (J-2)
Ann Arbor Transportation Authority

Number of Vehicles: 74 buses

Strategy
TransPlan is a program of the Ann Arbor Transportation Authority of Ann Arbor, MI, that includes an employer information packet that is used to promote, educate, and recruit employers and their employees on the availability of public transit and rideshare programs. The packet includes information on all services the agency offers, employee surveys on commuting, and an invitation to conduct presentations about the service at worksites. The packet is continually updated and can be customized for specific employers.

Objectives
To educate employers and employees about the commuting options available to them.

To increase ridership.

Resources
The program requires three hours of staff time per week to maintain. Printing of new inserts for the packets costs $1,500 a year.

Implementation Time
One to two months

Results
The agency has recorded consistent gains in ridership as a result of the program.

When
1992 and continuing

Contact
Liz Nowland-Margolis
Manager of Community Relations
Ann Arbor Transportation Authority
2700 S. Industrial Hwy.
Ann Arbor, MI 48104
Tel: 734/677-3901
Fax: 734/973-6338
E-mail: liznm@theride.org

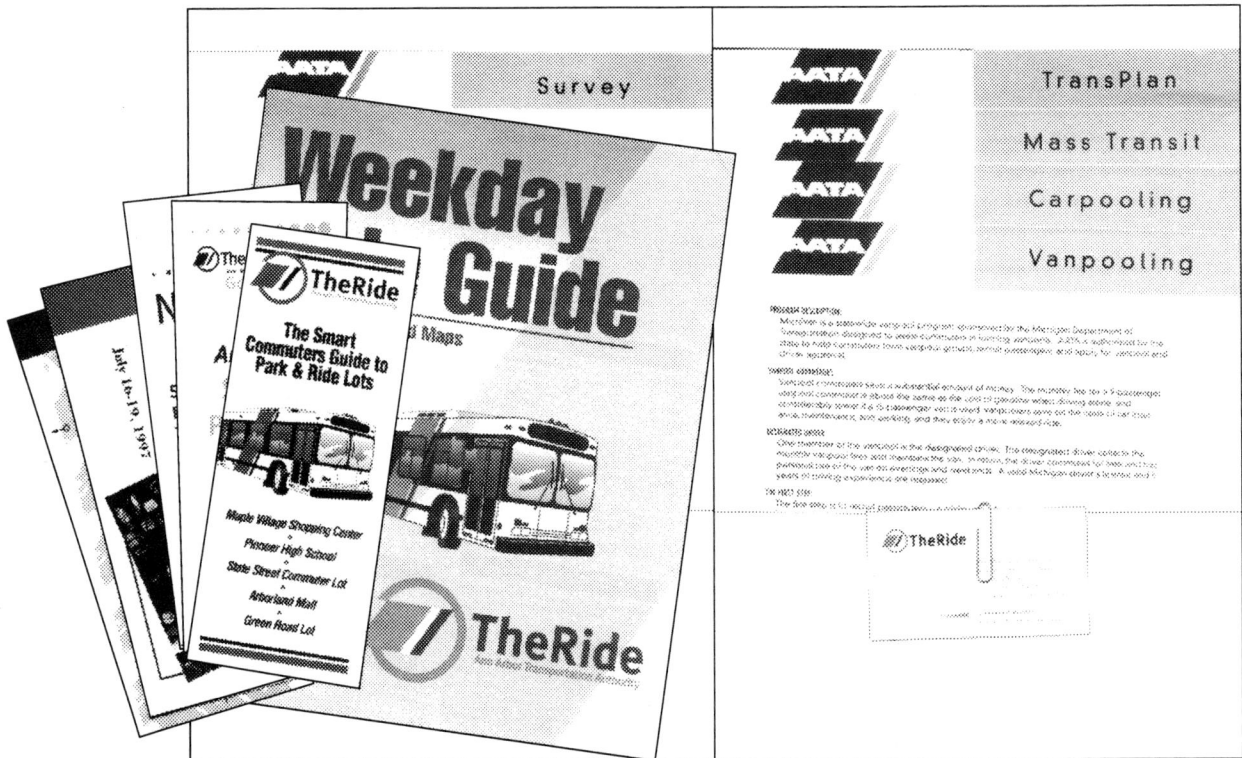

"Ride to Rewards" (J-3)   *Metro Transit*

Number of Vehicles: 900 buses

Strategy

In June 1995, Metro Transit of Minneapolis, MN, developed a frequent rider program that rewarded bus riders with incentives and rewards for regular use of its system. The "Ride to Rewards" program encourages more frequent use of Metro Transit services, while allowing the agency to create a customer database to build long-term relationships with its riders and provide for direct communication. Metro Transit believes more direct communication with its customers can effectively promote new services and lessen the impact of service changes. The agency is constantly seeking new business partners to provide incentives to customers registered in the database.

Objectives

To build long-lasting relationships with Metro Transit customers.

To create a database of customer information to allow for more direct communication.

Resources

Metro Transit developed a test program to measure customer reaction to incentives and rewards for increasing use of the transit system. The annual budget for maintaining the program is $35,000, of which almost half is spent on postage. The number of sponsors providing incentives varies with each mail-out.

Implementation Time

Four months

Results

The "Ride to Rewards" program customer database contains over 17,000 names.

When

June 1995 and continuing

Contact

Kathy Laudenslager
Market Development Specialist
Metro Transit
560 Sixth Av. North
Minneapolis, MN 55411
Tel: 612/349-7531
Fax: 612/349-7675
E-mail:
Kathy.Laudenslager
@metc.state.mn.us

Regional Guaranteed Ride Home (J-4)
Metro Commuter Services

Strategy

Metro Commuter Services (MCS) of St. Paul, MN, along with the Minnesota Department of Transportation, created the Regional Guaranteed Ride Home program to benefit all commuters who use alternative transportation. Commuters must ride the bus, carpool, bike, or walk to work at least three days per week to qualify for two coupons good for a free ride home by bus or taxicab. The ride value of the coupon is $20. Customers fill-out a registration form and then have it verified by their employer. Within 10 working days, the customer receives the 2 coupons.

The program is a reward for commuters who use alternative transportation and offers peace of mind to those who must work late or have an emergency. It also provides an incentive for commuters to consider using alternative transportation. Information about the program is distributed through direct mailings, interior bus cards, agency newsletters, and local newspapers.

Objectives

To reward commuters who use alternative transportation.

To entice commuters to use alternative transportation.

Resources

MCS administers the program and provides reimbursement for ridesharers, walkers, and bicyclists. Bus providers pay reimbursement costs for transit system riders. The Metropolitan Council and bus providers pay for promotion of the service. Reimbursement costs for the promotion average $2,000 a month.

Implementation Time

Four months

Results

The initial goal of the program was to register 3,300 customers for the service. By the end of the first year, the number of registered participants was more than 8,000.

When

September 1996 and continuing

Contact

Patty Carlson
Senior TDM Program Administrator
Metro Commuter Services
230 E. 5th St.
St. Paul, MN 55101
Tel: 651/602-1211
Fax: 651/602-1200
E-mail: patty.carlson @metc.state.mn.us

Rack Cards Promoting Destinations (J-5)

Greater Portland Transit District

Number of Vehicles: 24 buses

Strategy

To position the Greater Portland Transit District (METRO) of Portland, ME, as a valuable resource to its community, the agency developed a series of rack cards that promoted the diverse range of destinations within the City of Portland. The destinations included cultural facilities, social service and health agencies, recreational locations, and retail establishments. The agency used a targeted approach for the campaign, identifying organizations and merchants that would provide high promotional visibility for METRO. Copy for the rack cards was jointly developed with the organizations, focusing primarily on the benefits and services of the organizations. Two-thirds of a rack card was devoted to the destination and one-third promoted the METRO service. The cards were distributed to customers through location counter displays, direct mail, new member enclosures and registrations, and informational counters.

Objectives

To create multiple partnerships with targeted, highly visible organizations in the community.

To increase awareness of public transportation and position METRO as an important resource in the community.

Resources

METRO absorbed the cost of printing the rack cards, usually between $200 and $600 per card in quantities of 2,000 to 10,000 cards. Design and layout for the cards cost $100-$200. In some instances, the organization paid for the artwork for its card.

Implementation Time

One month per card

Results

The agency was able to alter its traditional marketing expenditures through a targeted, cost-effective approach. METRO continuously updates the rack cards and pursues new organizations in the community.

When

1995 and continuing

Contact

Philip Chin
Director of Marketing
Greater Portland Transit District
114 Valley St.
P.O. Box 1097
Portland, ME 04104-1097
Tel: 207/774-0351
Fax: 207/774-6241

Saturday Service in Saline County (J-6)
Rides Mass Transit District

Number of Vehicles: 45 vans, 3 buses

Strategy

In early 1997, Rides Mass Transit District of Rosiclare, IL, noticed that ridership on its route to the town of Harrisburg and two nearby towns in Saline County was beginning to decrease. Even though Harrisburg and the two other towns were small, the agency felt it was very important to provide service to the locations. The core ridership on the route was very loyal and considered the service to be a necessity. In an effort to boost ridership along the route, the agency developed a promotion that allowed free return trips on Saturdays with a coupon. The agency placed the coupon in local newspapers and on flyers distributed within the service area.

Objectives

To increase general public ridership on Saturdays in Saline County.

Resources

The promotion was conducted with very little cost to the agency. The flyers were developed in-house by the agency.

Implementation Time

One month

Results

In the two months of the promotion, ridership on the route increased 53 percent. When the promotion ended, ridership decreased but not to the level before the campaign, retaining some of the new riders.

When

February and March 1997

Contact

Betty S. Green
Executive Director
Rides Mass Transit District
P.O. Box 190
Rosiclare, IL 62982
Tel: 618/285-3342
Fax: 618/285-3340

RIDES MASS TRANSIT DISTRICT
SALINE COUNTY
SATURDAY SERVICE
252-4662

Saturday Service Schedule
8:00 A. M. - 2:00 P.M.

Need A Ride??

Time	Destination
8:00	Harrisburg
8:30	Carrier Mills
9:00 - 10:30	Harrisburg
10:00 - 10:30	Eldorado
11:15	Harrisburg
11:00 - 11:30	Carrier Mills
12:30 - 1:30	Harrisburg
1:00 - 1:30	Eldorado
2:00	Harrisburg

Call 252-4662 to make your Transportation Reservations

RMTD coupon

RMTD coupon

RMTD coupon   **MARCH SPECIAL**

Return trip Free every Saturday during the month of March with this coupon.

NAME _____

Red Carpet Saturdays (J-7)

Citrus Connection

Number of Vehicles: 30 buses, 13 mini-buses

Strategy

Red Carpet Saturdays is a special service that allows Citrus Connection of Lakeland, FL, to drive its buses up to two blocks off its regular fixed routes to pick up or deliver passengers at the door of their origin and destination. On Saturdays, the agency uses the same schedule on its fixed routes as weekdays. However, since there is less traffic on the roads and a lower number of passengers, buses have more time available to run the regular routes. This extra time allows bus operators to provide the door-to-door service and still maintain their schedule. Good coordination between the agency dispatcher and the drivers is essential to arrange the pick-up or delivery. The project was implemented through press releases, radio ads, and the agency's newsletter.

Objectives

To use existing resources to provide more valuable service at no additional cost to passengers.

Resources

There is no extra cost in providing the service. Marginal cost to the agency is negligible.

Implementation Time

Two to three weeks

Results

The promotion is considered to be very successful. Bus operators began to use the term "Red Carpet" as a verb, as in, "Can I have approval to 'Red Carpet' a passenger to the door of xyz shopping center?"

When

1989 and continuing

Contact

Steven Githens
Transit Director
Lakeland Area Mass Transit District
1212 George Jenkins Blvd.
Lakeland, FL 33815
Tel: 941/688-7433
Fax: 941/683-4132
E-mail: CitConGit@aol.com

Metro Service Awareness Campaign (J-8)
Niagara Frontier Transportation Authority

Number of Vehicles: 322 buses, 18 paratransit, 27 light rail vehicles

Strategy
In order to lessen the impact of a change in service, the Niagara Frontier Transportation Authority (Metro) in Buffalo, NY, offered a special weekly pass valid the week before and the first week after the change in service. The weekly pass contained several value-added features including discounts, free admission to local venues, and five free minutes of long distance calling. The agency maintained the value-added features in the next sales period of regular monthly passes.

Objectives
To build awareness of the level of Metro service available, thus stabilizing ridership.

To instill a sense of security in Metro customers that service will remain stable in the face of changes.

To encourage service trials by new riders through a weekly pass promotion.

Resources
The total cost of the campaign was $57,650, including printing, radio spots, and the free long distance calling offer. Eighty percent of the total cost was spent on the radio ads. Metro estimated a media trade-out value of $40,000 and the value of promotional features obtained through cooperative advertising at $50,000. One hundred hours of staff time was required to create the promotion.

Implementation Time
Two months

When
December 1995 to February 1996

Contact
Robert Gower
Acting Manager of Business Development
Niagara Frontier Transportation Authority
181 Ellicott St.
Buffalo, NY 14203-2298
Tel: 716/855-7646
Fax: 716/855-6387

SWIPER Promotion (J-9)

Capital District Transportation Authority

Number of Vehicles: 225 buses

Strategy

The Capital District Transportation Authority (CDTA) in Albany, NY, initiated a large-scale marketing campaign to introduce its new magnetic stripe monthly pass program, SWIPER. The new pass was advertised through television and print ads, radio spots, direct mail, and grass-roots sales calls.

Objectives

To introduce the new product and stimulate sales to individual and corporate customers.

Resources

The advertisements were produced by a contracted ad agency. The budget for the first year of the campaign was $150,000. In subsequent years, the promotion has required a budget of $75,000. Two staff members are required to maintain the program.

Implementation Time

Six months

Results

The old monthly pass program averaged 1,000 passes sold per month. The new SWIPER program averages 2,000 to 2,500 a month.

When

April 1994 and continuing

Contact

Carm Basile
Director of Marketing and Information
CDTA
110 Watervliet Avenue
Albany, NY 12206
Tel: 518/482-3371
Fax: 518/446-0675

"Extra Punch" Promotion (J-10)

Red Rose Transit Authority

Number of Vehicles: 45 buses, 33 vans

Strategy

The Red Rose Transit Authority (RRTA) in Lancaster, PA, conducts a promotion once a year in which customers who purchase an RRTA Ten Trip Ticket receive 11 trips for the price of 10. The "Extra Punch" promotion is advertised through interior bus signs, general mailings, and newspaper ads. Special versions of the tickets are printed and sold during the promotion.

Objectives

To convert cash fare riders to pre-paid transit users.

To boost sales of ten trip tickets.

To thank regular customers for their business. To involve bus drivers in promoting RRTA.

Resources

The promotion is low-cost. Printing cost for the tickets is approximately $250. Promotional materials are created by the agency.

Implementation Time

Three to four weeks per year

Results

RRTA believes the program is very successful. Regular customers wait for the promotion to stock-up on tickets. New pre-paid transit riders receive a bonus for making the change. During the promotion, sales of ten trip tickets increase 50 to 70 percent over normal monthly sales.

When

Conducted annually in February

Contact

Jennifer Burkhart
Marketing Manager
Red Rose Transit
Authority
45 Erick Road
Lancaster, PA 17601
Tel: 717/397-5613
Fax: 717/397-4761

Seasonal Promotions

easonal activities offer an excellent opportunity to promote a transit agency and the services it provides. Few other times of the year can create and attract as much transit ridership as holidays. As a result, the transit system is on display.

Events that attract a large number of citizens create a demand for the use of transit, especially if held in areas that contain limited access or parking. The community perceives increased use of the transit agency as lessening the potential for accidents and injuries since fewer drivers are in their own cars. By providing increased public safety during holidays, especially ones traditionally celebrated with alcohol such as New Year's Eve, the transit agency is seen as providing a public good and its image is enhanced.

A well-organized, efficient transit program during a holiday season or weekend can reap many rewards for an agency.

First Night Festivities Service (K-1)   *CT Transit*

Number of Vehicles: 379 buses

Strategy
Connecticut Transit (CT Transit) provides free bus service to area residents on New Year's Eve from 5:00 p.m. until 1:15 a.m. as part of First Night festivities in downtown Hartford, CT. The service

was provided in cooperation with the Hartford Downtown Council in order to encourage residents to attend the event. CT Transit placed ads in local newspapers promoting the service, and produced interior notices as well as press releases to inform the public.

Objectives
To create a positive co-op promotion with the Hartford Downtown Council. To increase ridership during the festivities.

To enhance the image of the transit agency in the community.

Resources
The total budget for this program was $6,365. This amount was broken down as follows: $1,500 for media buy, $2,200 for additional bus service, $2,500 for free fares value, and $165 for interior notices. The budget

has remained stable through the years. One employee in the marketing department works on the project.

Implementation Time

Two days of staff time.

Results

Free fares were provided for the first time in 1995 as part of the project. Ridership increased 800 percent over the previous year's festivities. As a result of providing the service, CT Transit received the event's "Grand Sponsorship" status ($5,000 value) with a free tag line on all of the event promotional materials, mention of the agency in all news releases and news conferences, sponsorship of one venue, as well as positive press coverage in several local newspapers. The service is provided every year and continues to be considered a success.

Adaptations

CT Transit uses similar promotions during Hartford's Festival of Jazz.

When

Service has been provided since 1990, with free fares first offered in 1995. The service is provided annually.

Contact

Maria McEvoy
Marketing Administrator
CT Transit
100 Leibert Road
PO Box 66
Hartford, CT 06141-0066
Tel: 860/522-8101 x312
Fax: 860/247-1810
E-mail: mmcevoy@cttransit.com

Holiday Lights Tour (K-2)   *Jackson Transit Authority*

Number of Vehicles: 13 buses, 4 paratransit vans

Strategy

Every year, on the two Saturdays before Christmas, Jackson Transit Authority of Jackson, TN offers a free tour of holiday lights displays to residents of senior centers and nursing homes. The tours take them through downtown and historic districts. The agency mails out information about the program to the homes and centers inviting their participation. Reservations by centers and homes are accepted on a "first call, first choice" basis for day and time.

Objectives

To enhance the image of the transit agency.

To enhance good public relations with the community.

Resources

There is no formal budget for this program. It requires one bus and operator. The driver is paid overtime for the tour service.

Implementation Time

Three months

Results

The program is considered a huge success. It is now in its ninth year. Many people that would not normally see the holiday lights are able to with this service. Several centers and homes call for reservations before the list is formally open.

When

Ongoing since 1989

Contact

Michele T. Jackson
Marketing/Planning
Coordinator
Jackson Transit Authority
241 E. Deaderick St.
PO Box 102
Jackson, TN 38301
Tel: 901/423-0200
Fax: 901/424-9323
E-mail: jtacjax@usit.net

Christmas "Stuff-a-Bus" (K-3)
Northwestern Connecticut Transit District

Number of Vehicles: 16 buses

Strategy

In collaboration with several civic agencies and organizations, Northwestern Connecticut Transit District (NW Conn Transit) in Torrington, CT decorates one of the buses as part of a holiday toy collection drive. The bus is decorated and parked at several local shopping centers on specific days. Shoppers are urged to tour the bus and donate a toy for a needy child. A money donation box is also made available. Several local businesses donate items for a free raffle at various sites. Coffee and doughnuts are served at some of the locations.

Objectives

To provide as many toys as possible for the needy children of the area.

To enhance the image of the transit agency as an organization that cares about the community.

Resources

The project was part of an interagency cooperation that included the Mayor's Office and the Torrington Firefighters Union. It was conducted by volunteers from transit staff, city hall staff, elementary and high school students, merchants, seniors, and political organizations. Agency staff time was approximately 50 person-hours. Local newspapers and cable stations were aware of the effort and the local papers ran a story on the project almost every day.

Implementation Time

One month

Results

The project is considered a major success. NW Conn Transit collected more than 2,000 toys and $1,500 in donations during the two-week program.

When

1996 and conducted annually.

Contact

Carol L. Deane
Office Manager
NW Conn Transit District
140 Main St.
Torrington, CT 06790
Tel: 860/489-2535
Fax: 860/489-3353
E-mail:
CDeane8127@aol.com

Northwestern Connecticut Transit District

Municipal Building
140 Main Street
Torrington, CT 06790
(203) 489-2535

PUBLIC SERVICE ANNOUNCEMENT

The Northwestern Connecticut Transit District, Mayor's Office, Torrington Fire Fighters Union, Mayor's Committee With Persons With Disabilities, Sullivan Senior Center and several City of Torrington employees are combining their efforts this year to promote "CHRISTMAS FOR CHILDREN".

"CHRISTMAS FOR CHILDREN" was set up to provide toys to needy children in the Torrington Area for Christmas. Our goal this year is to receive 1000 or more gifts. Kevin Hayes from the Torrington Fire Fighters Union will be distributing the toys through the FISH organization. In order to reach this goal the Transit District is going to create the "CHRISTMAS BUS". The Christmas bus will be provided by the Transit District, decorated appropriately and parked at random throughout the City of Torrington starting Sunday, December 1, 1996 through Sunday December 15, 1996.

Part of these Christmas festivities will feature many surprises for the people who come to tour the "CHRISTMAS BUS" and, hopefully leave a toy or two. One of the surprises will be a free raffle ticket to each of them with gifts being donated throughout the city.

Toys will also be accepted at the Northwestern Connecticut Transit District and the Torrington Fire Department on Water Street. Donations are also welcome. If your gift is in the form of a donation, checks should be made out to "CHRISTMAS FOR CHILDREN".

The "CHRISTMAS BUS" will be decorated at the Sullivan Senior Center on Wednesday, November 27 at approximately 10:30 AM. Volunteers are welcome. Anyone wishing to volunteer a couple hours of their time during this two week period can call Carol Deane at 489-2535.

Anything you can do to make this toy drive a huge success would be very much appreciated.

Any questions or suggestions can be directed to Carol Deane or Paul Tyrrell at the Northwestern Connecticut Transit District 489-2535 or Kevin Hayes, Torrington Fire Department, 489-2255.

Special
Events

Special Events

pecial events enable transit systems to participate in community efforts or to provide transit services to community activities. Transit systems may also choose to create events to familiarize infrequent transit riders with available services. Special events may provide good anchors for other targeted activities. The special event allows personal contact with a large number of people in a concentrated time period and is invaluable in launching or culminating advertising or promotional campaigns.

State and county fairs provide the same marketing opportunities as any special event does. Every state has a fair, often requiring effective transit services for success. For transit systems not located near the annual state fair, the same ideas and efforts can be applied to county and regional fairs.

Special Events Service (L-1)   *Maryland Mass Transit Administration*

Number of Vehicles: 819 buses, 100 rail cars, 35 light rail

Strategy

The Maryland Mass Transit Administration (MTA) has found that one of its most effective marketing strategies is the promotion of transit services to special events in the Baltimore area. Ridership to baseball and football games, fairground events, downtown events, and area festivals is always high. The primary strategy to promote service for special events is to utilize mass media (newspapers, radio). It is felt to be the most cost-effective method to reach potential riders. Another important strategy is to co-promote with the event organizers. For example, the MTA works closely with the Baltimore Orioles and Baltimore Ravens sports teams. An "MTA Service to Baltimore Ravens Games" brochure is included with the team's football season ticket holder mailing. Maryland MTA cross-promotes with every event they advertise. Events are targeted where parking is not free or plentiful, large numbers of attendees are expected, and the venue is very near mass transit options.

Objectives

To encourage MTA ridership to and from the special event that is being promoted. To entice trial of the system so that if the MTA service is convenient to the rider's home and work place, then the special events rider may convert to frequent MTA commuter use. It is believed that a commuter is more comfortable learning to use the system on the way to a special event than on the way to work.

Resources

Over $125,000 was budgeted in fiscal year 1997 for special events advertising. One staff member in the marketing division handles advertising, which accounts for 35 percent of his or her time. Another staff member works on promotions, primarily cross-promotions, which is approximately 60 percent of his or her time. Another staff member devotes 10 percent of his or her time on cross-promotions as liaison with the sports teams. For drivers and staff on event day, volunteers from within MTA are recruited and rewarded with compensatory or overtime pay.

Implementation Time

Service is ongoing and evolves accordingly.

Results

Many of the special events occur annually, so the service is repeated. While it is difficult to determine if these special events riders are converting to daily commuters on the system, nevertheless it is considered the best chance of reaching potential riders and familiarizing them with the system.

When

Annually

Contact

Buddy Alves
Acting Marketing Manager
Maryland Mass Transit Administration
6 Saint Paul Street
Baltimore, MD 21202
Tel: 410/767-8750
Fax: 410/333-3289
E-mail: GXLE71A@prod.com

Off-Peak Promotion (L-2)　　　　　　　　　*NJ Transit*

Number of Vehicles:
1,916 buses, 24 street cars, 780 rail cars

Strategy
To increase ridership during off-peak hours, New Jersey (NJ) Transit, the state-wide public transit system in New Jersey, has joined with several Manhattan entertainment venues in a free ride strategy. The venues, such as Madison Square Garden, Radio City Music Hall, and Broadway theaters, purchase a block of transit tickets from NJ Transit and then offer a NJ Transit free ride package with the purchase of performance tickets scheduled during off-peak hours. The package has been identified as the "NJ Transit Holiday Package" or "NJ Transit Christmas Package," and is part of the agency's Winter Services program. When purchased by a customer at the venue box office, Ticketmaster, or Tele-Charge, a transit ticket is included with each performance ticket. The customer is content because they have

received free transportation to the show, the venue has sold a seat to the performance, and NJ Transit has filled a seat on a bus or train not otherwise utilized.

Objectives
To increase off-peak ridership through linkage with special events.

Resources
Project budgets for the different versions of the program ranged from $8,000-$15,000 depending on the co-promotional agreement with the venues.

Implementation Time
Planning begins in June and amount of effort increases with each passing month.

Results
Package sales have increased form the previous year for each year the program has been offered. NJ Transit

attributes an increase in regular ridership to the program.

Adaptations
The agency also has package deals with their Summer Services program.

When
1994 and continuing

Contact
Jonathan Benjamin
Manager, Marketing Promotions
NJ Transit
1 Penn Plaza East
Newark, NJ 07105
Tel: 973/491-7148
Fax: 973/491-7567

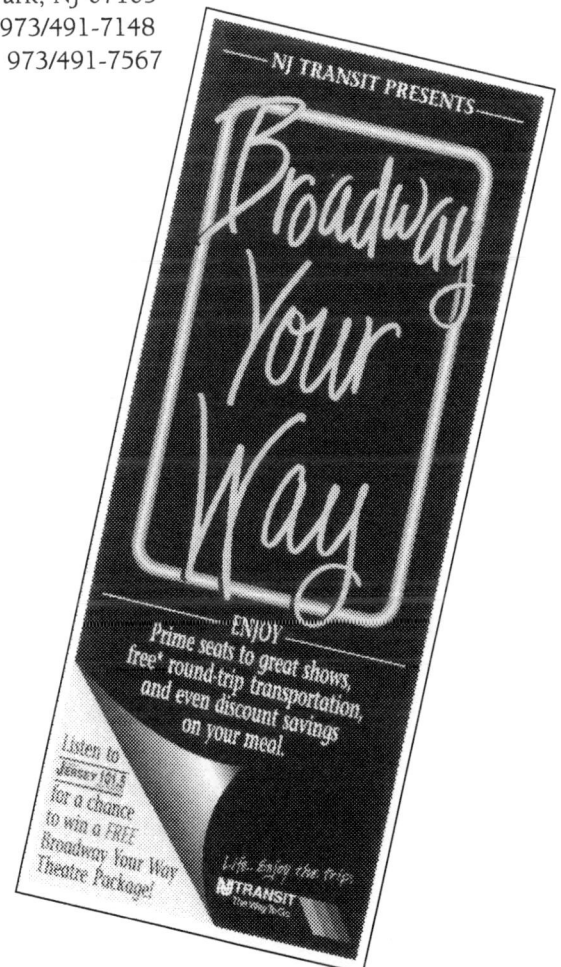

Earth Day Celebration (L-3)
Sacramento Regional Transit District

Number of Vehicles: 210 buses, 36 light rail vehicles

Strategy

The Sacramento Regional Transit District (RT) in Sacramento, CA transported more than 14,000 students and teachers, using RT's "Class Pass," to the annual Earth Day Celebration at the State Capitol. RT's "Class Pass" allows 10 or more students (grades 1-12) to travel for $1 dollar each with unlimited rides during non-peak hours. Adults traveling with the group pay only $2.50 each.

The RT's primary method of advertising the special fares were flyers included in informational packets sent to local elementary school teachers describing Earth Day Celebration events. The RT requires a two-week advance reservation for use of the pass, which allows RT time to coordinate buses and trains for the trips and prevent overbooking on the system. Teachers are directed to call RT's Customer Relations Call Center, which enters the reservation into their computer system that tracks group travel plans.

RT believes that teamwork and cooperation between departments is the key to making Earth Day events a success. Pre-event meetings with representatives from the Bus Transportation, Light Rail, Customer Relations, and Information Services departments take place early in the year. The representatives constitute the system's Earth Day Committee. The goals of the meetings are to keep all departments informed regarding planned activities and to fine-tune operating procedures based on previous years experience.

Objectives

To transport several thousand schoolchildren and adults to the State Capitol for Earth Day Celebration activities.

Resources

Agency personnel involved were Customer Service Representatives, marketing staff, bus and light rail operators, dispatchers, and supervisors. The budget for the project is under $5,000 and requires 150 person-hours.

Implementation Time

Three months

Results

The program is repeated each year. Ridership has increased every year of the event. The students and teachers are transported following the morning rush hour and returned before the afternoon peak period. Coordination of transport and personnel is considered very smooth due to good planning and previous experience.

Adaptations

The "Class Pass" is available throughout the year for school field trips and any group travel.

When

1992 and conducted annually

Contact

Jo Teele Noble/Ed Scofield
Marketing Representatives
Sacramento Regional Transit District
1400 29th St.
Sacramento, CA 95812
Tel: 916/321-2863
Fax: 916/444-0502

State Fair Traditions (L-4)

Metro Transit

Number of Vehicles: 900 buses

Strategy
Attending the Minnesota State Fair in Minneapolis is a yearly custom for many families. As a result, the theme for the advertising campaign for the 1996 State Fair became "State Fair Traditions," encouraging customers to make using Metro Transit part of their annual experience. All printed materials incorporated historical photographs provided by the Minnesota State Fair archives. Metro Transit, in partnership with the Minnesota State Fair, gave a $2 discount on gate admission to customers riding Metro Transit buses, providing an extra incentive to ride the bus.

Inserts were distributed to State Fair customers who purchased advance tickets from the middle of June through the beginning of the fair in August. Inserts were also mailed to a database of 16,000 frequent bus customers. Metro Transit produced 250,000 brochures and distributed them to businesses around boarding areas, sales outlets, schedule distribution sites, and State Fair outlets statewide. Interior bus cards and stories in TAKEOUT, Metro Transit's monthly onboard customers newsletter, distributed the State Fair message to current transit customers. To accommodate the large number of requests for telephone information, a special extension of the agency's automated schedule information line gave menu-driven information to callers.

On the return trip from the fair, Metro Transit handed out 50,000 brochures that thanked customers for riding the bus and included free ride coupons for future trips. Randomly selected customers were asked to complete surveys on service quality and willingness to ride again.

Objectives
To reduce traffic congestion around the fairgrounds.

To encourage customers to make riding the bus part of their annual State Fair experience. To illustrate the ease of riding transit and to encourage future commuting once familiarity has been established.

Resources
The budget for the project was $20,000. The State Fair provided advertising and distribution of materials.

Implementation Time
Eight months

Results
In 1996, one out of every eight people who attended the State Fair took a Metro Transit bus. The State Fair service created an extra 270,000 bus rides during the 12 days of the fair. This was a 9.2 percent increase from 1995 State Fair ridership and a 22 percent increase over 1992 figures.

Twenty percent of the coupons distributed to customers on the return trip from the fair were redeemed between the end of the fair and their expiration 60 days later.

Adaptations
Metro Transit has used the park-and-ride concept to provide service for Aquatennial summer events in 1998.

When
Since 1992, with service provided annually.

Contact
Kathy Laudenslager
Market Development Specialist
Metro Transit
560 Sixth Av. North
Minneapolis, MN 55411
Tel: 612/349-7531
Fax: 612/349-7675
E-mail:
Kathy.Laudenslager@metc.state.mn.us

Jackson County Fair Park 'N' Ride (L-5)
Rogue Valley Transportation District

Number of Vehicles: 26 buses

Strategy
Rogue Valley Transportation District (RVTD) of Medford, OR promoted their Park 'N' Ride service for the county fair through a coupon book exclusive to riders on RVTD buses attending the fair. The coupon book was for purchases at the local mall, Wal-Mart, and Eagle Hardware, which were the locations used as park 'n ride service lots. Twenty thousand coupon books were produced.

Objectives
To increase ridership on RVTD Park 'N' Ride service.

To enhance RVTD image as part of the local community.

To promote local business through a coupon book.

To mitigate traffic congestion during the week of the County Fair.

Resources
The Rogue Valley Mall, Wal-Mart, and the stores around them, absorbed the costs of producing the coupon book. Salvation Army members and other civic groups stuffed the coupon book into the agency's county fair packet in exchange for advertising on the buses. Local radio stations promoted the park 'n ride service and coupon book at no cost. One staff member of RVTD worked on the project. An RVTD envelope, which included the coupon book and several other individual coupons and was placed in the county fair packet, cost $900. The envelope had the logos of all the major sponsors and the County Fair printed on it. All other costs were underwritten by the other principals in the program.

Implementation Time
One and a half months

Results
The agency has noted an increase in ridership. The merchants are now doing direct advertising with the agency.

When
1997

Contact
Richard Smith
TDM Information/ Marketing
Rogue Valley Transportation District
3200 Crater Lake Av.
Medford, OR 97504-9075
Tel: 541/608-2420
Fax: 541/773-2877

Mid-State Fair Express Bus Service (L-6)
San Luis Obispo Regional Transit Authority

Number of Vehicles: 13 buses

Strategy
In order to help a segment of the outlying population attend the Mid-State Fair, the San Luis Obispo Regional Transit Authority (SLORTA) of San Luis Obispo, CA created an express bus service between several towns in the area and the main gate of the fair. The concern was that transit dependent segments of the population, such as the elderly, youth, and those without automobiles, would be unable to attend the fair because the local transit operations end at 6:00 p.m. SLORTA was able to provide this service through solicitation of corporate sponsorship by the fair from Chevron and a local television station, KSBY. Advertisements for the service were placed in the local television, radio, and print media. A marketing partnership with those entities kept costs low.

Objectives
To reduce traffic congestion on the county's roadways for the two-week period of the fair.

To provide a transportation option for transit-dependent citizens.

To provide low-cost transportation service to local residents and tourists.

To capture new transit users by exposing them to a convenient transportation option.

Resources
SLORTA uses four buses to provide three round trips a night for an 11-day period. Sponsorships are solicited from local television, radio, and newspaper media through trade agreements to provide advertising for the service. In addition, major sponsors, such as Chevron and Nextel, and the Fair itself are solicited for financial contributions. Partial funding is received through an Air Resources grant, with remaining costs absorbed through passenger fares and the SLORTA budget. Staff volunteers are positioned at passenger park and ride sites to answer questions about the service and guide passengers onto the buses.

Implementation Time
In the first year, three weeks were required to establish contacts and develop the service. Subsequent years have required one to two weeks to update the project.

Results
SLORTA's Mid-State Fair Express service has proven to be very successful in relieving traffic congestion through the Highway 101 corridor and around the fairgrounds. Ridership on the express service has increased 154 percent since its inception in 1995. In 1997, 6,250 passengers were transported.

Adaptations
SLORTA is considering the implementation of similar service to major events such as the Fourth of July celebration and an annual car festival in Pismo Beach.

When
1995 and conducted annually since.

Contact
Shari Presnall
Transit Systems Coordinator
San Luis Obispo Regional Transit Authority
1150 Osos St., Suite 206
San Luis Obispo, CA 93401
Tel: 805/781-4465
Fax: 805/781-1291
E-mail:
slorta@rideshare.org

Target Group Promotions

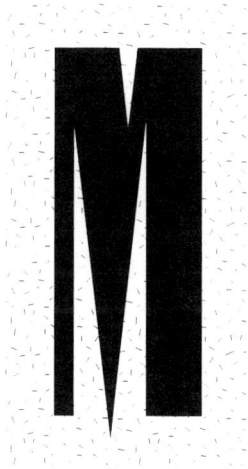

Many groups can be targeted in promotions by a transit agency to boost ridership, reward current users of the system, entice new ridership, and to educate segments of the population in the use and value of public transportation. Groups that are included in the examples below include: senior citizens, new employees, schoolchildren, high schools, libraries, and residents adjacent to bus routes.

"Behind the Wheel" Campaign (M-1)

Athens Transit

Number of Vehicles: 23 buses, 4 vans

Strategy
As part of elementary school class presentations and field trips, Athens Transit (The Bus) in Athens, GA takes photos of students sitting in the driver's seat of an agency bus. With parental approval, some of the photos are enlarged and used on interior bus cards that promote the transit agency.

Objectives
To increase awareness and raise enthusiasm for field trips to The Bus.

Resources
Film and development are the only additional costs to the agency. Enlargement of the photos is produced in-house on a color copier.

Implementation Time
One week

Results
Teachers call for copies of the photos and encourage other groups to schedule a field trip to the agency.

Parents often ride the bus to see their child's photo.

Adaptations
The promotion is used as part of Try Transit Week activities by Athens Transit.

When
1997 and continuing

Contact
Judy Dudley
Assistant Director
Athens Transit
325 Pound St.
Athens, GA 30601
Tel: 706/613-3432
Fax: 706/613-3433

Don't be Afraid!

You won't get lost!

Try Transit!

Transit Pass for Employees (M-2)

Number of Vehicles: 43 buses

Strategy

The Ames Transit Agency (CyRide) in Ames, IA conducts a new employee orientation and distributes transit passes for all new employees at Iowa State University (ISU). Agency staff give a five-minute presentation at university employee orientation programs in conjunction with departments at the university. After the presentations, free one-semester transit passes are distributed along with an offer of half-price passes for all subsequent semesters. The focus of the presentations is on new employees that will have difficulty securing good parking spaces on campus.

Objectives

To increase ridership of Iowa State University employees.

Resources

Two hours of staff time per month is required for the orientation. CyRide donates the first semester pass.

Implementation Time

One day

Results

The program is in its third year and will continue as long as ISU subsidizes the pass. The program averages approximately 150 employees per semester using the pass.

Adaptations

CyRide makes a presentation about its services to new students and their parents during student orientation.

When

1995 and continuing

Contact

Bob Bourne
Transit Director
CyRide
1700 W. 6th St.
Ames, IA 50010
Tel: 515/239-5563
Fax: 515/239-5578

"Riding Isn't" Campaign (M-2)

CyRide

Number of Vehicles: 43 buses

Strategy
To promote the ease of riding the bus for college students, the Ames Transit Agency (CyRide) in Ames, IA utilized a newspaper ad campaign that made a humorous comparison between things that are difficult or impossible and riding the bus. CyRide felt that non-riders of the bus may perceive it as a difficult thing to do. The advertisements ran on a rotating schedule in the Iowa State University newspaper. The ads were designed to be lighthearted and fun to a younger audience. The ads were eye-catching due to the use of color and vertical placement on the edge of consecutive pages in the newspaper.

Objectives
To increase college student ridership.

To promote the ease of riding transit.

Resources
The development of the campaign was done in-house. A single run of the ads in the student newspaper costs approximately $1,200.

Implementation Time
One to two days

Results
The promotion heightened the awareness of the service available and increased ridership. Focus groups comprised of the target audience were used to determine whether the objectives had been met. According to the focus groups, nearly 75 percent remembered seeing the ads.

Many of them stated that the ads reinforced the value and ease of the bus service. Ridership increased 18.2 percent during the four-month period of the campaign.

The advertising campaign received a 1996 AdWheels Award for the use of print media by a small transit agency.

Adaptations
The agency has used a similar campaign for other target groups.

When
1995

Contact
Barbara Neal
Senior Operations Assistant
CyRide
1700 W. 6th St.
Ames, IA 50014
Tel: 515/239-5565
Fax: 515/239-5578

230 m.p.h. at Indy is treacherous ... *or try 300 m.p.h. at Pomona* — CyRide — Riding isn't!

0 to 100 in 4 seconds is brutal .... *and uses a lot of fuel* — CyRide — Riding isn't!

cornering a Porsche 911 is hard ... *unless you're a professional* — CyRide — Riding isn't!

Transit Brochure Distribution (M-4)

Number of Vehicles: 9 buses, 4 vans

Strategy

Rural Transit in Bloomington, IN informs current and potential riders of the services available to them by distributing brochures to businesses and agencies along Rural Transit's routes. The brochures were designed to be easy to read and understand. Permission to distribute the brochures is gained by contacting the manager of each agency and business. Businesses are also encouraged to inform their staff about the benefits of riding public transit.

Objectives

To increase awareness of the services provided by Rural Transit.

Resources

The project was funded through a special marketing grant of $2,000 from the Indiana Department of Transportation. An Indiana University student summer intern was integral to the creation and implementation of the campaign.

Implementation Time

One year

Results

Public awareness of Rural Transit services increased and good working relationships were established with the local businesses and agencies. Ridership increases were noted by the agency.

Adaptations

Rural Transit distributes an advertising sales brochure to local businesses.

When

1997 and continuing

Contact

Jewel Echelbarger
Executive Director
Rural Transit
7500 W. Reeves Rd.
Bloomington, IN 47404
Tel: 812/876-3383
Fax: 812/876-9922
E-mail: jechelba
@bloomington.in.us

"Summerdime Days" (M-5)

LAKETRAN

Number of Vehicles: 83 buses

Strategy

LAKETRAN in Grand River, OH sponsors an art contest for the design of the agency's "Summerdime Days" logo to be used for direct mailers, promotional tee-shirts, and newspaper ads. "Summerdime Days" allows all passengers to ride the agency's fixed-route system for 10 cents on weekdays between Memorial Day and Labor Day. The contest is open to all high school students in Lake County. Letters are mailed to each high school in the county listing the criteria for entries. A winner is selected at each high school, at their discretion, and receives $100 and is entered for the LAKETRAN Grand Prize. A panel of local artists selects the grand prize from the participating schools. The winner receives $500 and is announced at a LAKETRAN Board of Trustees meeting with the student's parents, art teacher, and principal in attendance. Articles about the contest appear in the local daily newspaper.

Objectives

To promote public transit to youth ages 12 to 18.

Resources

No special budget is required for the contest, except for the prize money. Required staff time is approximately 15 hours.

Implementation Time

Two months

Results

The contest is in its third year and the agency considers it to be highly successful.

When

1996 and conducted annually.

Contact

Frank J. Polivka
General Manager
LAKETRAN
PO Box 158
Grand River, OH 44045
Tel: 216/350-1000
Fax: 216/354-4202

"The RRTA Senior Game" (M-6)
Red Rose Transit Authority

Number of Vehicles: 45 buses, 33 vans

Strategy
Red Rose Transit Authority (RRTA) in Lancaster, PA conducts a six-week frequent rider

Play the 2nd Annual
Senior Game
Have the driver punch this card each time you board an RRTA bus. When all 4 blocks are punched, driver will collect card and enter it in drawings for weekly prizes and the $1,500 Grand Prize. See reverse for details.

Name:
Address:
City:
Zip: State:
          Phone:
Information required in order to win prizes.

№ 15401 | 1 | 2 | 3 | 4

promotion for senior citizens age 65 and over called "The RRTA Senior Game." Game cards are distributed by operators and punched each time a senior citizen uses the system. Four punches completes a card, which is then entered into a drawing for prizes. The more frequently seniors ride during the promotion, the more likely they are to receive prizes. Weekly drawings are held and the winner's names are posted inside buses and the agency's information center. All entries are eligible for a grand prize. The agency prefers many small prizes which create many winners, rather than a few big ones. The contest is promoted with a mailing to local senior citizen groups, ads in senior citizen publications, and interior bus ads.

Objectives
To induce senior citizens to try RRTA.

To reward current senior citizen riders.

Resources
The agency spends a total of $4,000 on the promotion, which is matched by a local mall in prizes. The special cards cost approximately $250 to print.

Implementation Time
Two to three weeks per year.

Results
Feedback from senior citizens participating in the contest was very positive. Ridership increased during the period of the promotion. In 1996, over 11,000 game cards were returned in a six-week period which represents almost 47,000 rides. Over a five-week promotional period in 1997, approximately 7,500 cards were submitted representing over 30,000 rides.

When
1995 and conducted annually in October and November.

Contact
Jennifer Burkhart
Marketing Manager
Red Rose Transit Authority
45 Erick Road
Lancaster, PA 17601
Tel: 717/397-5613
Fax: 717/397-4761

"Garage Sale Cruise" (M-7)   *Kosciusko Area Bus Service*

Number of Vehicles: 14 buses

Strategy
Kosciusko Area Bus Service (KABS) in Warsaw, IN takes residents of a local senior center on a three-hour "Garage Sale Cruise" once a week during the summer months. The service is advertised by a series of posters and notices delivered to the senior center. Word-of-mouth advertising is very strong with this promotion.

Objectives
To create a core ridership at the senior center that may then use KABS for their everyday transportation needs.

Resources
Reserve vehicles are used for the service. Three hours of staff time per week during the summer is required. The cost to the agency is minimal.

Implementation Time
One to two months

Results
Senior citizens at the centers are very supportive of the program. Several public transportation advocates have grown from the core group.

When
Every summer since 1995.

Contact
Rita Baker
Transportation Coordinator
Kosciusko Area Bus Service
1804 E. Winona
Warsaw, IN 46580
Tel: 219/267-4990
Fax: 219/267-4990
E-mail: KABS@KConLine.com

Flyers Distributed on the Virginia Tech Campus (M-8)
Blacksburg Transit

Number of Vehicles: 30 buses, 5 vans

Strategy
Blacksburg Transit in Blacksburg, VA posts single page flyers throughout the Virginia Tech campus promoting its paratransit service. The flyers are placed in and around several main buildings, especially the health services and athletic buildings. The agency uses crutches as a symbol on the flyer, rather than a wheelchair, because 90 percent of the students using the paratransit service are temporarily disabled, mainly through injuries.

Objectives
To increase awareness of the agency's paratransit service on campus.

Resources
Design and printing of the flyers is done by the agency. Twelve hours of staff time is required, including distribution. The cost of the advertising campaign is minimal.

Implementation Time
Two months

Results
Calls to the agency requesting information about the service increased. Applications for the service increased by 350 percent.

When
1996 and continuing

Contact
Kevan Danker
Paratransit Coordinator
Blacksburg Transit
2800 Commerce St.
Blacksburg, VA 24060-6656
Tel: 540/961-1185
Fax: 540/951-3142
E-mail: blacksburg.
paratransit@bev.net

Bus Service Guide in Hotels near an Airport (M-9)

Pace Suburban Bus

Number of Vehicles: 638 buses, 374 paratransit vehicles, 321 vanpools

Strategy

Pace Suburban Bus in Arlington Heights, IL distributes the Rosemont Area Bus Guide to 21 hotels located directly east of O'Hare International Airport, one of the busiest in the world. The agency felt that the hotels in Rosemont, IL offered a lucrative market in which to promote its services. Hotel employees, travelers, and tourists were targeted with the brochure. Rosemont contains a major convention center, a major theater, and an indoor arena. The area is well-served by Pace and the agency felt the target groups could be provided with alternative destinations other than downtown Chicago. Agency staff reviewed other materials distributed to the hotels to determine what would attract a potential user's attention.

Objectives

To target potential markets in the agency's service area.

To promote new ridership.

Resources

Approximately 24 hours of staff time was required for the program. Agency staff determined the information and destinations included in the brochure. The agency's graphics department designed and printed the brochure. A brochure distribution service costs $900 annually.

Implementation Time

Three months

Results

A five percent increase in ridership was recorded when the brochure was first distributed. The agency has received favorable comments from several of the hotels, and the brochure is regularly reprinted in order to replenish stock.

Adaptations

Pace Suburban Bus also distributes guides pertaining to entertainment destinations in its service area, special events service, and other communities. The guides are distributed to major employers and other hotels.

When

1995 and continuing

Contact

Barbara Ladner
Manager of Market Development
Pace Suburban Bus
550 W. Algonquin Rd.
Arlington Heights, IL 60005
Tel: 847/228-2467
Fax: 847/956-7916
E-mail: barbara.ladner@pacebus.com

"The Transit Connection - Connecting the Worker to the Workplace" (M-10)
Triangle Transit Authority

Number of Vehicles: 24 buses, 50 vans

Strategy

The Triangle Transit Authority (TTA) in Research Triangle Park, NC held three pairs of job fairs, six in total, that focused on the importance of public transit options for the workplace.

The fairs were conducted in response to the State of North Carolina's welfare reform initiative and attempted to match low-income residents with area employers. Transportation to and from places of employment is a major issue for these residents and the fairs are an opportunity for the agency to become informed of the needs of both employers and applicants and to inform both groups about the services offered by TTA.

The job fairs were held in pairs so TTA could utilize the same advertising for both. The fairs were advertised through local media, public service resources, and through cooperative efforts with the employers. Flyers were also distributed throughout the local communities and included a coupon which offered a free ride to the fair.

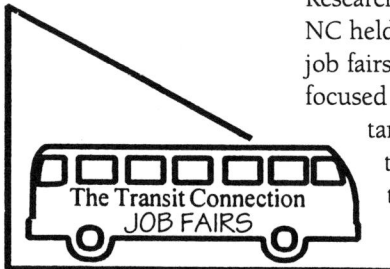

"Connecting the Worker to the Workplace"

Objectives

To bring employers and job seekers together for their mutual benefit.

To educate both groups about the benefits and availability of public transportation options.

To increase ridership.

To be responsive to job-related transportation needs by working with employers and other community groups to find creative transportation solutions for the workplace.

Resources

Approximately $15,000 was spent by TTA on each pair of job fairs. Five hundred hours of staff time was required to implement the program.

Implementation Time

Nine months

Results

TTA considers the project to have been a great success. Workers and employers were brought together and many of those hired are using public transportation. Several employers now consult TTA about their employee transportation needs. Strong relationships with community organizations and the business community have been established and enhanced. The third and fourth job fairs were conducted in response to requests by the participating groups. TTA is now represented on several area agency boards for work-related transportation issues, and the agency is asked to make presentations to different groups offering programs and seminars on welfare-to-work issues. TTA's involvement in regional welfare reform programs has become a continuous role for the agency.

When

1996 and continuing

Contact

Billie Cox
Transit Marketing and Rideshare Manager
Triangle Transit Authority
PO Box 13787
Research Triangle Park, NC 27709
Tel: 919/990-9038, x 20
Fax: 919/990-9127
E-mail: b.cox@mail.gte.net

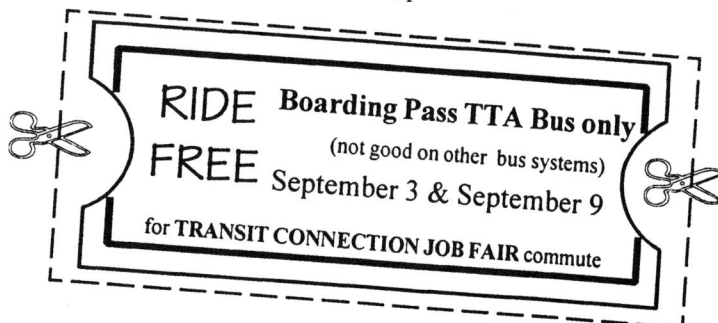

"Class Pass" (M-11)

Number of Vehicles: 56 buses

Strategy

The "Class Pass" program enables teachers at schools in the service area of the Eastern Contra Costa Transit Authority (Tri Delta Transit) in Antioch, CA to take their classes on field trips for free. The trips are taken on regularly planned routes at the scheduled times. Teachers inform the agency of the trip requirements and the agency plans the trip for them. They are limited to two field trips on the system every school year. Tri Delta Transit also offers a presentation to teach children about bus safety and rules. The agency shows a video and distributes a bag of transit marketing "goodies," including "toy bus" cookies, a bus bank, a transit coloring book and crayons, a Tri Delta Tran-

sit schedule and various school supplies. The service is promoted by sending letters containing a poster and brochure to the superintendents of the school districts and mailing a brochure to every teacher at schools in the agency's service area. Word-of-mouth among the teachers has also helped to promote the service.

Objectives

To teach children how to use public transit.

To distribute agency schedules and maps to homes in the service area.

To build ridership.

Resources

Approximately 1.5 hours of staff time are required per field trip. The gift bags cost $3 per student.

Implementation Time

Three months

Results

The service has been a major success for Tri Delta Transit. Teachers regularly ask for the service. Student ridership has increased since the program's inception.

When

1996

Contact

Deborah Bass
Marketing Coordinator
Tri Delta Transit
801 Wilbur Av.
Antioch, CA 94509
Tel: 925/754-6622
Fax: 925/757-2530
E-mail: dbass@eccta.org

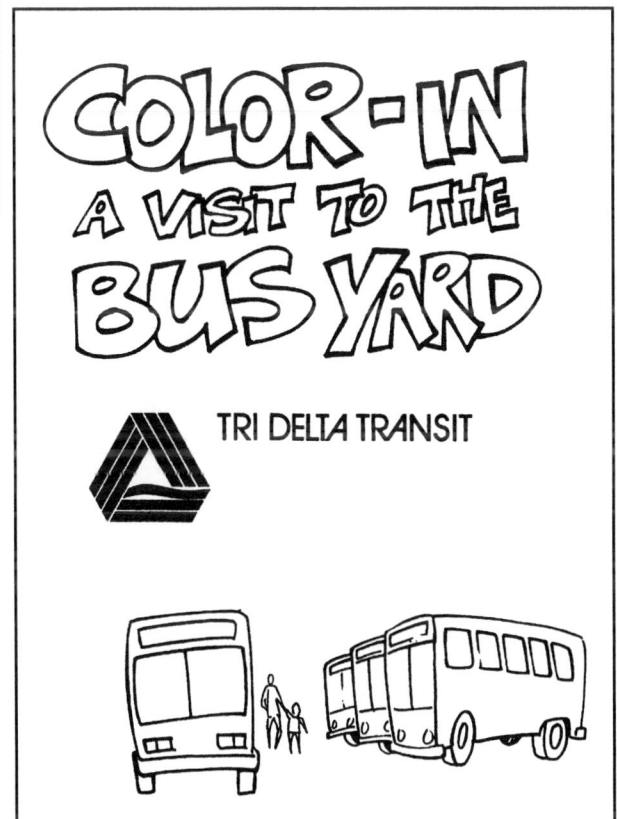

"Smart Tripper Book" and Tour (M-12) — *Kitsap Transit*

Number of Vehicles: 167 buses, 45 paratransit buses, 94 vans

Strategy

Kitsap Transit in Bremerton, WA developed a student handbook and tour, geared to 6th graders, in order to introduce students to the problems of air pollution, traffic congestion, and energy consumption, and to promote the use of alternative transportation as part of the solution to these problems. Students study the Smart Tripper Book in class, then visit Kitsap Transit for a tour of the facility. The agency works with the school district, offering the book and tour free of charge as part of the school year curriculum. The focus was placed on 6th graders because the agency felt they were old enough to be conscious of transportation and environmental issues.

Objectives

To develop future transit riders and transit supporters.

Resources

The annual cost of the program is approximately $14,500. Seventeen tours are conducted during the school year. Approximately $2,000 is required annually to produce the Smart Tripper Book. Each tour requires a total of 28 hours of staff time.

Implementation Time

Four months

Results

Teacher evaluations of the book and tour after participating rate the program as "excellent." Student pass sales and student ID card requests at the agency have increased since the program started. Kitsap Transit was awarded a grant from the Washington State Energy Office in 1996 to revise the Smart Tripper Book to make it generic enough for all Washington state transit systems. The book was printed and distributed to all transit agencies and school districts in the state.

Adaptations

The book is updated periodically and the tour is evaluated and revised as needed each year.

When

1993 and continuing

Contact

Bob Ferguson
Transportation Demand Management Administrator
Kitsap Transit
234 S. Wycoff Av.
Bremerton, WA 98312-4199
Tel: 360/478-5864
Fax: 360/377-7086

"Get On Board" (M-13)
Erie Metropolitan Transit Authority

Number of Vehicles: 55 buses, 1 trolley

Strategy
Erie Metropolitan Transit Authority (EMTA) in Erie, PA conducts a multi-faceted transit awareness program called "Get On Board." The agency holds transit awareness assemblies in each of the local elementary schools with information about the system and what it accomplishes for the community. Coloring books and search-a-word sheets are distributed to the children and education lessons are given to the teachers. The students are asked to complete a coloring sheet about "where they would like to go on the bus" for a chance to win prizes such as EMTA plaques or items from area merchants. Classes are asked to complete a bulletin board about transit and submit a picture of their creation to the agency. The class with the most informative and creative bulletin board wins a trip on the EMTA Trolley to

the agency for a tour of its facilities, EMTA tee-shirts, and a class plaque. The winning class is also allowed to name a bus for an entire year.

The program receives free advertising from sponsors on a local radio station as a community tie-in. The sponsors donate prizes and radio personalities attend the assemblies. The program also receives occasional coverage in the local newspaper.

Objectives
To educate schoolchildren on the value and use of public transit in the community.

Resources
EMTA spends $3,000 a year to maintain the program, primarily for copying, printing, and stickers. Free advertising is garnered from a local radio station. Prizes are donated by advertisers on the radio station.

Implementation Time
Three to four months

Results
In the first year of the program, 10 out of 14 local elementary schools

GET ON BOARD

were involved. Good working relationships were developed with sponsors and the radio station. The number of schools involved continues to grow.

When
1996 and continuing

Contact
Alyson Amendola
Marketing Director
EMTA
PO Box 2057
127 E. 14th St.
Erie, PA 16512-2057
Tel: 814/459-8922
Fax: 814/456-9032

School Poster Contest (M-14)   *Virginia Railway Express*

Number of Vehicles: 67 railcars, 13 locomotives

Strategy
Virginia Railway Express (VRE) in Alexandria, VA sponsors an annual poster contest for fourth

and fifth graders in its service area. The contest is coordinated through the school district. It is designed to coincide with the fourth grade Virginia History and fifth grade Colonial Life studies in their respective curriculums. The winning class is awarded a field trip to an historic part of the City of Alexandria aboard VRE. The top three entries are reproduced and posted in every VRE train and station for four months.

Objectives
To interest schoolchildren, their parents, and teachers in using the train system.

To help promote the agency's 50 percent fare discount for school groups.

Resources
Two VRE staff members accompany the field trip. The tour guides at the destination cost $170, while the chartered bus that returns the children in the afternoon costs the agency $300. The cost for printing the winning posters is approximately $3,000.

Implementation Time
Five months

Results
Over 300 entries are received on average. The class field trip garners newspaper coverage. Positive relations are maintained with the local school districts.

When
1996 and conducted annually.

Contact
Ann King
Manager of Marketing and Customer Service
Virginia Railway Express
6800 Versar Center, Suite 247
Springfield, VA 22151
Tel: 703/684-1001
Fax: 703/684-1313
E-mail: gotrains@vre.org

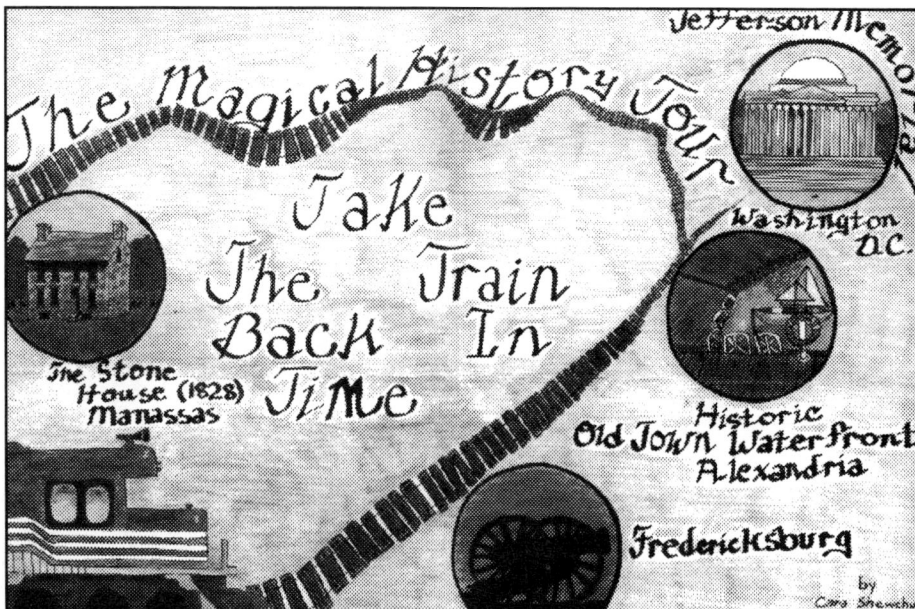

"Books On Board" (M-15)

St. Cloud Metropolitan Transit Commission

Number of Vehicles: 31 buses, 12 paratransit vehicles

Strategy

In 1989, the St. Cloud Metropolitan Transit Commission (Metro Bus) in St. Cloud, MN began a program with support from the local newspaper, the St. Cloud Times, called "Books On Board." Elementary school classes received free tokens from the agency, rode the system on its regular routes, and toured area businesses. Upon returning to their classrooms, the students were instructed to write and illustrate a book, individually or as a class, about their trip. The finished products were displayed on Metro Buses.

Teacher information packets are distributed to elementary schools in the fall semester. Teachers arrange their own tour date and time from a provided list of participating businesses. Agency staff visit schools and make presentations before the scheduled trip to teach the schoolchildren about Metro Bus. Pencils and bus-shaped erasers are distributed to the students. A $50 savings bond is awarded to a student in each participating school through a random drawing.

Objectives

To teach children about public transportation.

To teach children the process of creating a book.

Resources

Required staff time to implement the program is 30-40 hours. The agency spends approximately $850 a year to maintain the program.

Implementation Time

Two to three months

Results

Over 1,200 children participated in the program during the latest school year. The agency receives very positive publicity for the program.

When

1989 and ongoing

Contact

Kim McCarney
Administrative Assistant
St. Cloud Metropolitan
Transit Commission
665 Franklin Av. NE
St. Cloud, MN 56304
Tel: 320/251-1499
Fax: 320/251-3499

"The Pass" (M-16)

Tri-Met

Number of Vehicles: 766 buses, 26 light rail vehicles

Strategy

The Tri-County Metropolitan Transportation District of Oregon (Tri-Met) in Portland provides "The Pass," a discounted summer pass for youth 18 years and under that allows unlimited riding on Tri-Met from June 1 through August 31. The present cost of a pass is $40. Five area merchants are recruited each year to provide a special discount to pass holders. Merchant logos and discounts are printed on the back of the pass. Usually, merchants with large numbers of local outlets are chosen to facilitate student use of the discounts. The merchants must also agree to display advertising posters for "The Pass."

Objectives

To maintain school year ridership levels during the summer months.

To build long-term relationships with the target market.

Resources

The total budget per year for the summer youth pass program is approximately $50,000. The agency spends approximately $34,000 on advertising including print, radio, television, posters, brochures distributed to area high schools, and bus interior and exterior ads.

Implementation Time

Five months per year

Results

In the first year "The Pass" was offered, a 200 percent sales increase over the previous summer sales period was recorded. Sales have increased an average of five percent in succeeding years.

When

1994 and conducted annually in the summer months.

Contact

Beth Erlendson
Marketing Representative
Tri-Met
4012 Southeast 17th Ave.
Portland, OR 97202-3993
Tel: 503/239-6438
Fax: 503/239-6469
E-mail: erlendsb@tri-met.org

Summer Youth Bus Pass (M-17)   *SLORTA*

Number of Vehicles: 13 buses

Strategy
The San Luis Obispo Regional Transit Authority (SLORTA) in San Luis Obispo, CA offers a summer youth bus pass that provides 15 weeks of unlimited rides on 4 fixed-route systems in San Luis Obispo County. The cost of the pass is $15, which provides the slogan "15 Weeks for 15 Bucks." SLORTA contacted the transit agencies and negotiated their agreement to accept a universal summer youth pass and share in the revenue on a percentage basis. The Air Pollution Control District - County of San Luis Obispo provides sponsorship for ad costs due to the trip reduction as a result of the program. The pass is advertised on local radio and television stations and in the area newspaper.

Objectives
To teach youths how to use public transit.

To increase ridership during the summer months.

To reduce the number of vehicle trips by parents and other family drivers.

To provide youths in rural areas with access to recreational activities not found in their local area.

To distribute route and schedule information to parents.

Resources
The total cost of the program is $10,000. Half of the total goes towards advertising, the other half for additional schedules and maps.

Implementation Time
One week per year

Results
SLORTA has recorded increases in pass sales for every year of the program. The pass provides good benefits for kids in rural areas of the county who otherwise would not have ready access to area recreational activities.

When
1994 and conducted annually during the summer.

Contact
Shari Presnall
Transit Systems Coordinator

San Luis Obispo Regional Transit Authority
1150 Osos St., Suite 206
San Luis Obispo, CA 93401
Tel: 805/781-4465
Fax: 805/781-1291
E-mail: slorta@rideshare.org

"Easy Rider" Student Pass Program (M-18) — *Palm Tran*

Number of Vehicles: 156 buses, 55 vans

Strategy

The "Easy Rider" Student Pass Program conducted by the Palm Beach County Surface Transportation Department (Palm Tran)

in West Palm Beach, FL has two components: a monthly program during the school year and a summer program. The agency tries to increase sales of student passes during the summer by making their use fun. Coupons from merchants or organizations that target the youth market are added to the pass which, if redeemed, make the pass free. Palm Tran has also held contests in the form of a game card to encourage participation in the pass program. Children were also allowed to name the program for the agency.

The pass program is marketed through presentations at local schools and ads in student newspapers. Partnerships are secured with local merchants to co-promote the pass and provide discounts. Radio, television, and newspaper advertising is utilized by Palm Tran to promote the pass.

Objectives

To increase student ridership.

To build relationships with the local business community.

Resources

Advertising for the "Easy Rider" pass program is part of the annual expenditures of the agency's marketing department. Three merchants are chosen to provide the coupons.

Implementation Time

Eight weeks

Results

Sales increases of the student passes have been recorded by Palm Tran. The first year of the program saw a 400 percent increase in pass sales from the previous year. The program is successful enough to inspire local merchants to contact the agency about providing coupons with the passes.

When

1996 and continuing

Contact

Toni Sanders
Marketing Director
Palm Tran
1440 PBIA
West Palm Beach, FL 33406
Tel: 561/266-9507
Fax: 561/266-9498

"Library Pass-Time" (M-19)   *Valley Transit*

Number of Vehicles: 28 buses

Strategy

On Wednesdays throughout the summer months, anyone under 18 with a local library card can ride free on Valley Transit in Appleton, WI. The program is called "Library Pass-time" and is co-promoted with a local radio station and the local newspaper. Advertising runs for the duration of the promotion. Posters touting the program are displayed on buses, at the central transit center, and at the libraries. Bus headsigns and letters to the schools are also used by Valley Transit to promote the offer.

Objectives

To increase ridership during the summer months.

To promote summer reading by students.

To enhance relationships with the community.

Resources

Newspaper ads, radio spots, and poster designs are created internally by the agency. The cost of printing the posters is approximately $100. Fifty 10-ride tickets, tee-shirts, frisbees, and other promotional items are given to the 5 local libraries to be offered as prizes or giveaways. Required staff time to maintain the program is 10-12 hours.

Implementation Time

One week

Results

The local libraries have provided good responses to the promotion. Ridership is higher on Wednesdays during the summer. Many of the children utilizing the offer are usually accompanied by a parent paying full fare.

Adaptations

Follow-up discussions are held with each library in order to improve the program for the next summer.

When

1990 and continuing

Contact

Jennie Eastman-Kiesow
Marketing Coordinator
Valley Transit
801 Whitman Av.
Appleton, WI 54914
Tel: 920/832-6100
Fax: 920/832-1631

"Ride 'n' Read" Program (M-20)

Muncie Public Transportation Corporation

Number of Vehicles: 23 buses, 13 paratransit vehicles

Strategy

Muncie Public Transportation Corporation in Muncie, IN conducts the "Ride 'n' Read" promotion, which is a cooperative program with the city's libraries. The agency provides bus tickets to the libraries good for a ride home and a ride back to the library. When children check-out books, a librarian issues him or her a bus ticket. Fliers distributed at schools and libraries are used to promote the campaign.

Objectives

To encourage trial ridership by youth.

To encourage reading by students.

Resources

Very little staff time is required to implement and maintain the program. The program costs approximately $400 to maintain.

Implementation Time

One to two months

Results

The program has exceeded the agency's expectations. Ridership has steadily increased since the promotion began. Teachers have used the tickets in conjunction with field trips to the library.

When

1991 and continuing

Contact

Mary Gaston
Assistant General Manager
Muncie Public Transportation Corporation
1300 E. Seymour St.
Muncie, IN 47302
Tel: 765/282-2762
Fax: 765/287-2385

Metro Benefits Direct Mail Campaign (M-21)

Niagara Frontier Transportation Authority

Number of Vehicles: 322 buses, 18 paratransit, 27 light rail vehicles

Strategy

Individuals living within three-quarters of a mile of six bus routes were targeted by Niagara Frontier Transportation Authority (Metro) in Buffalo, NY for a direct mail campaign touting the agency's services. The direct mail targeted over 20,640 residents in census blocks in the three-quarter mile "buffer zone" along the routes containing significant percentages of individuals with lifestyles congruent with rider profiles from previous market research. Addresses were obtained from the agency's in-house database of its service region.

The direct mail piece consisted of a folded flat identifying the route with its positive features highlighted. The flat contained a return postcard designed to accommodate additional information requests. Each direct mail flyer received print media support advertisements inserted in publications distributed in targeted areas during the week of the mailing.

Objectives

To target specific demographic segments within a three-quarter mile buffer zone along Metro Bus routes.

To increase buffer zone resident awareness of Metro Bus service.

To induce ridership among the buffer zone residents.

To develop a rider database using a return mail postcard in the direct mail flat.

Resources

The total cost of the campaign was approximately $10,000. These costs included the flyers, print media ads, postage and handling. Thirty hours of agency staff time was required for the campaign.

Implementation Time

Four months

Results

Results of the campaign were measured by comparing ridership as measured by the farebox for a period of four weeks prior to the promotion, during the week of the promotion, and for four weeks after the mail-out. Three of the six routes saw modest gains of one-to-three percent. The other 3 routes saw increases from 11 to 33 percent. A total of 443 requests for service information were received. Metro's direct mail campaign received first place in the 1995 APTA AdWheel Awards for "shoestring" campaigns conducted by large transit systems.

When

1994-1995

Contact

Robert Gower
Acting Manager of Business Development
Niagara Frontier Transportation Authority
181 Ellicott St.
Buffalo, NY 14203-2298
Tel: 716/855-7646
Fax: 716/855-6387

Direct Mail Piece (M-22)   *Cottonwood Area Transit System*

Number of Vehicles: 3 mini-buses

Strategy

The Cottonwood Area Transit System (CATS) in Cottonwood, AZ mailed a one-page flyer to over 7,000 residential and business addresses in its service area. The flyer provided detailed information about the services offered by the agency. The flyers were all in a hot pink color and, when folded, the agency logo appeared on the front.

Objectives

To increase ridership through awareness of service availability.

Resources

The transit agency spent approximately $1,500 on the mail-out, which included the costs of material, mailing permit, and postage. One of the cost-saving techniques used by CATS was the hiring of individuals with disabilities from a local association to perform tasks such as folding, stapling, and stamping the flyers.

Implementation Time

One month

Results

The flyer proved to be very useful for the agency and the local community. Demand for CATS service increased to the point that new vehicles needed to be purchased by the agency. Some customers were still referring to the flyer three years after receiving a copy.

When

1989

Contact

Joseph E. Paulus
Transportation Manager
Cottonwood Area Transit System
827 N. Main St.
Cottonwood, AZ 86326
Tel: 520/634-5526
Fax: 520/634-5520

Cottonwood Area Transit System

C.A.T.S

634-2287

(634-CATS)

Serving the Cottonwood, Clarkdale, Verde Village & Bridgeport Areas.

FARES:
IN COTTONWOOD CITY LIMITS: $1.50 ONE-WAY & $3.00 ROUND TRIP
ALL OTHER LOCATIONS: $2.00 ONE-WAY & $4.00 ROUND TRIP

DISCOUNTS: BUY A 20 - TRIP PASS! 20% DISCOUNT RATE
COTTONWOOD: $24.00 FOR 30 SERVICE VALUE ($1.20 EA. ONE-WAY TRIP)
ALL OTHERS: $32.00 FOR 40 SERVICE VALUE ($1.60 EA. ONE-WAY TRIP)
**Ride free on your Birthday!!! Offer good for one (1) round-trip on your Birthday, or, during the week of your Birthday. Must show driver verification of Birthday and ask driver for "Birthday Ticket."

•GOT THE "TRANSPORTATION BLUES?"
•WANT DOOR-TO-DOOR SERVICE?

•NEED A RIDE?
•GIVE US A CALL!

❑ Dispatcher available Monday through Friday from 8:00 a.m. to 5:00 p.m. All vehicles radio-dispatched. Three vehicles wheelchair-lift equipped (handicapped accessible service).

❑ Service schedule Monday through Friday from 7:00 a.m. to 5:00 p.m. and Saturdays 9:00 a.m. to 2:00 p.m. ★Please Note: No dispatcher available on Saturdays. Service is by advance reservation only.

❑ For best service, call & schedule your appointment 48 hours in advance.

❑ WORKING PARENTS WITH CHILDREN IN SCHOOL...KEEP US IN MIND FOR YOUR CHILDREN'S TRANSPORTATION NEEDS.

C.A.T.S DOES NOT OPERATE IN OBSERVANCE OF THE FOLLOWING HOLIDAYS:

New Year's Day (Jan. 1), President's Day (3rd Mon. in Feb.), Memorial Day (last Mon. in May), Independence Day (July 4), Labor Day (1st Mon. in Sept.), Veteran's Day (Nov. 11), Thanksgiving Day (4th Thurs. in Nov.), Day following Thanksgiving, Christmas Day (December 25).

Non-Transit Users Survey (M-23)

Fresno County Rural Transit Agency

Number of Vehicles: 30 buses

Strategy

As part of a tri-annual performance audit, Fresno County Rural Transit Agency (FCRTA) in Fresno, CA conducted a survey of non-transit users in its service area. FCRTA provides accessible, real-time, demand responsive intra-city service and scheduled fixed-route inter-city transit services linking rural communities in the county to each other and with the Fresno-Clovis Metropolitan Area. The agency decided to make the survey a marketing informational tool for lasting value. Special attention was given to develop a survey format that would prove inviting to the reader, provide enough inducement to contact FCRTA for services, and include an incentive to try the service. The survey included a service area map, agency phone numbers, and a free inter-city round trip pass. It was written in English and Spanish. Five thousand surveys were mailed at random to service area residents.

Objectives

To survey non-transit users to determine how agency rural public

transportation operations and marketing could be modified to increase ridership.

To enhance agency efforts to market its services more effectively to non-traditional transit dependent patrons.

Resources

The regional Council of Fresno County Governments acquired a Federal Transit Administration grant in order to assist FCRTA in conducting the survey. The total budget for the project was approximately $15,000.

Implementation Time

Six weeks

Results

The survey response rate was 2.1 percent, considered average for this type of campaign. Responding citizens who provided their name and address received a Fresno County Transportation Guide from the agency. Modest ridership increases were recorded following the survey.

Adaptations

In the summer of 1998, FCRTA conducted a quality service survey targeting senior citizens in its service area.

When

1996-1997

Contact

Jeffrey D. Webster
General Manager
Fresno County Rural
Transit Agency
2100 Tulare St., Suite 619
Fresno, CA 93721
Tel: 209/233-6789
Fax: 209/233-9645
E-mail:
fcrta@lightspeed.com

---

1 one — **New Rider Survey**

1. **How do you currently travel to and from your home?**
   - [ ] Your own car/truck/van
   - [ ] Ride with someone else
   - [ ] Motorcycle/Bicycle
   - [ ] Walk
   - [ ] Public Transit
   - [ ] Other: _____
   - [ ] No means of transportation

2. **If you were to use our services, where would you need to go?**
   - [ ] Work
   - [ ] School
   - [ ] Medical/Dental
   - [ ] Shopping/Errands
   - [ ] Social
   - [ ] Recreational
   - [ ] Senior Center
   - [ ] Other: _____

3. **How often would you travel with us?**
   - ____ Number of times per week
   - [ ] Only when needed

4. **How many people living in your household would travel with us?**
   - ____ Number of children
   - ____ Number of adults

5. **Do you have any disabilities that require special travel assistance?**
   - [ ] No
   - [ ] Yes, wheelchair, or other: _____

6. **Would you like information about carpooling?**
   - [ ] Yes. *(Please complete below.)*
   - [ ] No

7. **Would you like to receive a free copy of the Fresno County Transportation Guide detailing all the available services within Fresno County?**
   - [ ] Yes. *(Please complete below).*
   - [ ] No

Optional Section:
Name: _____
Address: _____
City, Zip: _____

Thank you for taking the time to complete our survey!

To go to **2**,
fold down the top portion of this page

Route-Specific Information Campaign (M-24)
GP Transit

Number of Vehicles: 49 buses

Strategy

Greater Peoria Mass Transit District (GP Transit) in Peoria, IL conducts an information campaign for residents and businesses along specific routes in its service area. Homes and businesses within one to two blocks of the route are targeted, each receiving an information packet. Each packet contains a route schedule, letter of introduction, special event information, coupons, a reply card, and a free bus ticket for the specific route. Over 4,000 packets are distributed either directly or placed on doorknobs.

Objectives

To increase awareness of agency services.

To increase ridership along specific routes.

Resources

An average packet distribution along a route costs approximately $2,000, including printing, compilation, and distribution.

Implementation Time

Two weeks per route.

Results

Ridership increases of two percent on the targeted routes are recorded by the agency in the month after distribution of information packets. It is an ongoing campaign and the agency is proceeding systematically through its entire system.

When

1997 and continuing

Contact

Donna Calvin
Director of Marketing/Public Relations
GP Transit
2105 NE Jefferson Av.
Peoria, IL 61603
Tel: 309/676-8015
Fax: 309/676-8373

Colma New Move-in Mailer (M-25) *Sam Trans*

Number of Vehicles: 345 buses

Strategy

Because of an extension of the Bay Area Rapid Transit (BART) system to Colma in North San Mateo County, several bus routes of the San Mateo County Transit District (Sam Trans) in San Carlos, CA were greatly altered or deleted. In attempt to increase ridership on SamTrans buses to the Colma BART station, the agency conducted a direct mail campaign to new residents in the northern portion of San Mateo County. The agency felt that new residents represented the best opportunity for ridership since their commute patterns had not yet been established. The direct mail packet contained 10 free, one-way bus tickets.

Objectives

To increase ridership on SamTrans routes to Colma BART station in North San Mateo County.

To increase ridership on BART.

Resources

The agency spent approximately $43,000 to produce the direct mail packets. The cost per packet was $4.

Implementation Time

One to two months

Results

An 11 percent increase in ridership in the northern parts of the county was attributable to the direct mail campaign. Increases in ridership on BART from the Colma station were also noted.

Adaptations

The agency used the same type of campaign in the southern part of San Mateo county in order to increase ridership along El Camino Real, a main corridor in the area.

When

February and April 1997

Contact

Gordon Smith
Marketing Manager
SamTrans
PO Box 3006
1250 San Carlos Av.
San Carlos, CA 94070-1306
Tel: 650/508-6249
Fax:650/508-6443
E-mail:
smithg@samtrans.com

Try Transit Week

Try Transit Week

ry Transit Week is an annual event initiated by the American Public Transit Association. The event is observed by public transit systems across the nation with a variety of activities, such as a day of reduced fares or no fares at some systems, educational presentations at schools and civic organizations, and distribution of transit marketing materials. Try Transit Week helps identify the benefits of using public transportation to current and potential riders, and often showcases a transit system's most creative marketing ideas.

"Try Transit Day" Promotion (N-1)
Southern Oklahoma Rural Transportation System

Number of Vehicles: 18 vans, 4 buses

Strategy
The Southern Oklahoma Rural Transportation System (SORTS) of Durant, OK, is a demand responsive transit agency serving a four-county area in southeastern Oklahoma. The agency offers a "Try Transit Day" promotion during Try Transit Week, allowing customers free rides throughout the day. The same type of promotion is used by SORTS during special community events in its service area. These events include a Senior Olympics, blood drives, and various conferences. The free rides are then offered to and from the events. The agency is able to garner free advertising for its offer by inclusion in the promotional materials of event organizers.

Objectives
To increase awareness of the role of transit in the community and its importance to the economy.

Resources
There is no cost to SORTS for the promotions. The agency believes any lost fares from the promotions are made up by new ridership.

Implementation Time
Three weeks per event

Results
The agency averages a 10 percent increase in ridership following the promotions.

When
Annually during Try Transit Week and for certain local community events.

Contact
Allen M. Leaird
Director of Transportation
Southern Oklahoma Rural Transportation System
P.O. Box 1577
Durant, OK 74702-1577
Tel: 580/924-5332
Fax: 580/920-2004
E-mail: bigfive@fullnet.net

CarFree Days of Spring (N-2)

Boise Urban Stages

Number of Vehicles: 36 buses

Strategy

CarFree Days of Spring was a five-day program conducted by Boise Urban Stages (The BUS) in Boise, ID, to promote the use and benefits of transit. The program was sponsored by the Mayor's Office and included an opening day press conference and transportation fair. In the course of the five days there were various promotions, including a Bike to Work Day, a Transit Appreciation Day (on which all rides were free), a scavenger hunt regarding transportation issues for high school students called High School Hunt, and Learn It All at City Hall in which citizens received answers to their transportation questions. Throughout the week, local businesses competed in the Great Employer Ride-Off. Companies of similar size competed to see which could achieve the highest percentage of employees using alternative transportation. The events culminated with the Mayor's Celebration Luncheon.

Objectives

To raise community awareness of transit.

To encourage trial ridership of the system.

To reward current users of the system.

Resources

The Mayor's Office and the transit agency helped to recruit businesses for the employer contest. The Mayor sent letters promoting the contest to each member of the Chamber of Commerce. Interested companies were contacted by agency staff who explained the contest in detail and provided a packet of information. The packet included sample posters, sample e-mails, ideas on how to promote the contest internally, fun facts, etc. The cost of the event to Boise Urban Stages was approximately $1,300 for printing, postage, and prizes.

Implementation Time

Three months

Results

The promotion was considered to be a great success. The agency measured increases in the amount of press coverage, citizen participation in individual events, and employer participation in the contest for each year of the event. On average, more than 2,000 single occupancy vehicle trips were eliminated during the event.

Adaptations

Many of the events, with the exception of the Great Employer Ride-Off, have been incorporated into the agency's annual Try Transit Week activities.

When

1993-1997

Contact

Kelli Fairless
Manager of Operations
Boise Urban Stages
300 S. Avenue A
Boise, ID 83702-6299
Tel: 208/336-1019
Fax: 208/336-9048

Elementary School Drawing Contest (N-3)
Ohio Department of Transportation

Strategy
As part of its preparation for Try Transit Week in 1997, the Ohio Department of Transportation (ODOT) sponsored a drawing contest for fifth graders at ODOT's Adopt-a-School in Columbus, OH. The students were asked to draw and color a scene depicting an aspect of public transportation. The entries were then displayed in ODOT's offices and employees were asked to vote for the best three. The top two entries were enlarged and used as Try Transit Week posters for display in various state buildings. Winning students also received ODOT certificates of appreciation.

Objectives
To enhance the positive image of public transportation in Ohio.

To educate children on the benefits of public transportation.

To advertise Try Transit Week to children, parents, ODOT, and other state agencies.

Resources
Approximately 22 hours of staff time was required to implement the contest. Entry forms, voting materials, and printing costs were approximately $200.

Implementation Time
Three months

Results
One-third of the fifth graders submitted entries. The winning entries were displayed in ODOT offices and various state buildings.

Adaptations
ODOT staff made more informative presentations to each participating class which has led to an increased participation rate of 80 percent.

When
1997 and repeated annually

Contact
Dave Seech
Planner
Ohio Department of Transportation
1980 West Broad St.
P.O. Box 0899
Columbus, OH 43223
Tel: 614/644-9515
Fax: 614/466-0822
E-mail:
dseech@odot.dot.gov

Try Transit Week '93 (N-4)

Number of Vehicles: 10 trolleys

Strategy

In the City of Charleston, SC, the Visitor Reception and Transportation Center serves as a hub for three Downtown Area Shuttle (DASH) routes. The center also served as the focus for Try Transit Week activities in 1993.

In the weeks leading up to the event, presentations were made to local businesses soliciting their support. Organizations responded by donating free passes to local attractions, discount coupons, and gift certificates to be used as prizes. The logos of the six largest sponsors were printed on the back of Try Transit Week '93 tee-shirts and in newspaper ads for the event.

A poster contest promoting the week's activities was conducted. The contest was open to all third- and fourth-graders in Charleston County. Winning posters were displayed at the transportation center throughout the weeklong event, and the winning student's class received tee-shirts and a field trip.

Public service announcements promoting Try Transit Week activities were broadcast on local television and cable channels during the two weeks leading up to the events. An opening day press conference at the center with local officials and poster contest winners was held. It received coverage on two local television stations and an article in the local newspaper.

Over the course of the week, several events and activities were conducted. Daily drawings for prizes were held, with trolley drivers distributing entry forms. A free fare day was offered, with city employees boarding the trolleys and distributing promotional items. Several school groups took advantage of the opportunity and used the DASH for their field trips. Throughout the week, city employees also distributed gifts to drivers that had been contributed by local businesses. On Employee Appreciation Day, local merchants sponsored a party for transit employees at a downtown restaurant. Try Transit Week '93 culminated with a skating party held at the transportation center for transit and city employees, poster contest winners and their families and classmates.

Objectives

To promote better understanding of the benefits of public transportation in the Charleston area.

To develop a good working relationship between the business community and the transit system.

To increase local ridership on the system.

Resources

The cost of the promotion was $4,000 to the DASH, supplemented by more than $8,500 in contributions donated by the local business community. The donations were primarily in the form of gifts and coupons. One full-time staff member devoted approximately 320 work-hours to develop and implement the project. One part-time employee spent 40 hours.

Implementation Time

Three months

Results

During the week of events, ridership increased by 20 percent. Productive working relationships were established between the agency and local businesses. DASH believes that the event helped change public perceptions - known as a system serving visitors to one meeting the needs of the general public.

When

May 1993

Contact

Susan B. Richards
President
SR Concepts
2718 Cordwainer Court
Charleston, SC 29414
Tel: 803/769-6159
Fax: 800/670-4737
E-mail:
rich0505@aol.com

Try METRO Week 1997 (N-5)

Number of Vehicles:
1,344 buses

Strategy
Try METRO Week is a seven-day event conducted by the Metropolitan Transit Authority of Harris County (METRO) in Houston, TX, and promotes the use of public transit in Houston. A special $5 pass is created that allows an individual to use METRO during the promotion week with unlimited trips on any bus route. Part of the promotion includes an annual survey of riders. As an incentive for mailing in the survey, a $750 gift certificate from Best Buy is awarded through a random drawing of names of those submitting the survey. Employees of METRO are saluted during the week with a free employee night at a major local multi-screen cinema. Employees also receive a commemorative tee-shirt.

The promotion was marketed through print and electronic media. Radio spots were aired during high traffic periods on multiple stations. Fliers, bus cards, decals, newsletters, and fax-based collateral were also used to promote the event. Promotional/sales booths were placed throughout the city advertising the special features available during the week.

Objectives
To enhance awareness of public transit in Houston.

To invite trial ridership of the system.

To reward current riders and employees.

Resources
Total expenditures by METRO for the event was more than $75,000, the majority of which was spent on television ads and tee-shirts.

Implementation Time
Two to three weeks

Results
Try METRO Week 1997 saw a 3.8 percent increase in ridership over the previous year. Seven percent of total ridership during the week was considered trial ridership. Sales of the special passes showed an increase over the previous year.

When
May 1997 and conducted annually

Contact
Traci Romero
Manager of Advertising and Promotions
METRO
1201 Louisiana, Room 20033
Houston, TX 77002
Tel: 713/739-4011
Fax: 713/739-3791

Free Ride Coupon and Letter to the Editor (N-6)

MARQ-TRAN

Number of Vehicles: 30 buses

Strategy

As part of its efforts for National Transportation Week, the Marquette County Transit Authority (MARQ-TRAN) in Marquette, MI, printed an advertisement in the local daily newspaper in a "letter to the editor" format. The letter provided facts about the transit agency and invited citizens to try the system during Transportation Week. Included in the ad was a coupon that could be redeemed for one free ride.

Objectives

To increase awareness of the transit system as part of National Transportation Week.

Resources

The letter was drafted by MARQ-TRAN staff and signed by the chairperson of the agency's board of directors. The ad cost $225.

Implementation Time

One day

Results

More than 750 coupons were redeemed over the course of the week. The agency believes 75 percent of them represented new riders.

Adaptations

This marketing technique has also proven effective during elections involving funding of the system.

When

1996

Contact

Howard Schweppe
Human Resources Officer
Marquette County Transit Authority
145 W. Spring St.
Marquette, MI 49855
Tel: 906/225-1283
Fax: 906/225-0682

marqtran
MARQUETTE COUNTY TRANSPORTATION SYSTEM

Dear Citizens of Marquette County:

May 13th through May 18th marks the celebration of Transportation Week in the State of Michigan. In Marquette County, the theme for this year's celebration is "The Future is Riding On Us."

Riding a bus may not seem like an act of freedom and independence but to many older adults and people with disabilities, it is. Public transportation plays an absolutely essential role in providing senior citizens and people with disabilities the mobility necessary to enjoy life to its fullest and also compete in the workplace to take full advantage of all the opportunities of substantive employment.

In our rugged climate, public transportation also provides an important alternate means of transportation for those who cannot, or prefer not to drive at night or in inclement weather.

The "shared ride" philosophy of public transit is also at the heart of the environmental and social issues facing our society. As our resources become more scarce and our environment more fragile, the role of public transportation becomes increasingly more prominent. Public transportation stands as a rational and practical alternative to increased individual transportation.

Here in Marquette County we are fortunate to have MARQ-TRAN, a 26-bus local public transit system which continues to grow and develop and play an active role in our county's economic development. Transportation is the common thread which ties together people, goods and services.

Since its inception thirteen years ago, MARQ-TRAN has grown and developed into the largest non-urban transportation provider in the State of Michigan. Last year MARQ-TRAN's six fixed route buses and five door-to-door buses traveled 813,142 miles. MARQ-TRAN's operational policies and procedures set the standard for all other transit systems to follow. We have dedicated employees who sincerely care about the people they serve and who are sensitive to special needs.

Transportation Week marks an annual recognition of the need for public transportation in our community. It also provides an opportunity to acknowledge the contributions made by public transit to our county's growth and development. As we look to the future, the continued dividends earned by an investment in public transit become more clear. The future IS riding on us!

Sincerely,

Lois J. Paquet
Chairperson, Board of Directors
Marquette County Transit Authority

P.S. On behalf of MARQ-TRAN's employees, I invite you to sample our fine transportation system free of charge. Clip the coupon attached below. It's good for a free ride on any MARQ-TRAN bus during Transportation Week.

FREE! FREE! FREE! FREE! FREE! FREE! FREE!
Good for ONE FREE RIDE on any MARQ-TRAN Bus during Transportation Week

CROSS
REFERENCES

CROSS
REFERENCES

Cross References of Project/Strategy Categories

0 - Primary Category
X - Secondary Categories

Code	Title of Project/Strategy	A - Accessibility-Related Projects	B - Community Events	C - Cooperative Promotions	D - Image Promotions	E - Internal Promotions	F - Introduction of New Service	G - Media Relations	H - Problem-Solving Projects	I - Promoting Transit	J - Rider Inducements	K - Seasonal Promotions	L - Special Events	M - Target Group Promotions	N - Try Transit Week
A-1	Radio Ad for Rider Training Program for People with Disabilities	0												X	
A-2	TRIP Center Directory	0													
A-3	Promoting Fixed-Route Service for Medical Trips	0								X				X	
B-1	Co-sponsor of a Clean Air Fair	X	0	X											
B-2	Person of the Year with Disabilities Award	X	0	X											
B-3	Clean Air Challenge		0							X				X	
B-4	"Bike the Bay" Campaign		0				X							X	
B-5	"Our Own Words" Poetry Contest		0	X				X						X	
B-6	Stuff-A-Bus Promotion		0	X	X			X				X			
C-1	"Here's the Scoop" Campaign			0					X						
C-2	The Talking Yellow Pages			0										X	
C-3	Rider's Guide Publication			0										X	
C-4	Co-op Program with the Knoxville Museum of Art			0									X		
C-5	"Sally Says" Placemats for McDonald's			0	X									X	
C-6	New Residents Program			0						X				X	
C-7	Co-op Grocery Store Program for Seniors			0					X					X	

O - Primary Category
X - Secondary Categories

Code	Title of Project/Strategy	A - Accessibility-Related Projects	B - Community Events	C - Cooperative Promotions	D - Image Promotions	E - Internal Promotions	F - Introduction of New Service	G - Media Relations	H - Problem-Solving Projects	I - Promoting Transit	J - Rider Inducements	K - Seasonal Promotions	L - Special Events	M - Target Group Promotions	N - Try Transit Week
C-8	Bus Book Advertising			O					X						
C-9	Painted Bus Program			O									X		
C-10	TRIP Employer Pass Program			O										X	
C-11	"Transit Works!"			O										X	
C-12	FoxTrot Promotion			O			X								
C-13	Rider Incentives			O							X				
C-14	United Shoppers Service	X		O										X	
C-15	Weekend Shopper's Shuttle			O						X					
D-1	Bus Naming Contest				O					X				X	X
D-2	Thumbody Express-ions				O				X	X					
D-3	Video for Speaker's Bureau	X			O					X					
D-4	"On the Move" Newsletter				O					X					
D-5	Changing Paint Schemes				O										
D-6	Star Trolleys				O					X					
D-7	METRO Online Website				O				X						
D-8	"Pledge to Our Customers" & Customer Service Tour				O	X			X						

O - Primary Category
X - Secondary Categories

Code	Title of Project/Strategy	A - Accessibility-Related Projects	B - Community Events	C - Cooperative Promotions	D - Image Promotions	E - Internal Promotions	F - Introduction of New Service	G - Media Relations	H - Problem-Solving Projects	I - Promoting Transit	J - Rider Inducements	K - Seasonal Promotions	L - Special Events	M - Target Group Promotions	N - Try Transit Week
D-9	Mural Beautification Project				O				X	X					
E-1	Blue Jeans for Needy Families				X	O									
E-2	Wellness Program					O									
E-3	Driver's Excellence Award		X		X	O									X
E-4	On-Site Child Development Center				X	O			X						
E-5	VIA for Life Health Fair					O									
F-1	Free Fun-Filled Fridays		X	X			O				X				
F-2	Trolley Fiesta		X	X			O								
F-3	Magic Bus Design Contest			X			O			X					
G-1	Radio Shows				X			O							
G-2	"Smile on Monday" Contest				X			O							
G-3	"The Big Wheel Contest"			X				O		X				X	
G-4	Media Trade-outs							O							
G-5	Media Bus Drivers							O		X				X	
H-1	Bus Line Promotion Postcard								O				X	X	
H-2	Promotion of Teleride						X		O					X	

O - Primary Category
X - Secondary Categories

Code	Title of Project/Strategy	A - Accessibility-Related Projects	B - Community Events	C - Cooperative Promotions	D - Image Promotions	E - Internal Promotions	F - Introduction of New Service	G - Media Relations	H - Problem-Solving Projects	I - Promoting Transit	J - Rider Inducements	K - Seasonal Promotions	L - Special Events	M - Target Group Promotions	N - Try Transit Week
H-3	Customer Behavior Program								O						
H-4	Bus Stop Blitz						X		O					X	
H-5	Park-N-Ride Campaign								O	X				X	
H-6	Communibus 316 Relaunch	X		X					O	X				X	
H-7	Mural Art in Transit			X	X				O						
H-8	"From the Driver's Seat"				X	X			O						
I-1	Transportation Fair Booth		X							O					
I-2	Cable Television Advertising									O					
I-3	State Capitol Public Transit Display				X					O					X
I-4	Advertising and Brochure Promoting Ridership				X					O					
I-5	"I Have Connections"				X					O	X				
I-6	"Found Time" Contest			X						O					
I-7	Annual Meeting					X				O					
I-8	"Step Off the Gas, Step On the Bus"									O	X				
I-9	"Those Lextran Drivers"				X	X				O					
J-1	Free Ride Wednesdays and Saturdays								X	O	O			X	X

O - Primary Category
X - Secondary Categories

Code	Title of Project/Strategy	A - Accessibility-Related Projects	B - Community Events	C - Cooperative Promotions	D - Image Promotions	E - Internal Promotions	F - Introduction of New Service	G - Media Relations	H - Problem-Solving Projects	I - Promoting Transit	J - Rider Inducements	K - Seasonal Promotions	L - Special Events	M - Target Group Promotions	N - Try Transit Week
J-2	TransPlan Employer Information Packet									X	O			X	
J-3	"Ride to Rewards"									X	O				
J-4	Regional Guaranteed Ride Home									X	O			X	
J-5	Rack Cards Promoting Destinations			X	X					X	O				
J-6	Saturday Service in Saline County								X		O				
J-7	Red Carpet Saturdays	X									O				
J-8	Metro Service Awareness Campaign				X				X	X	O			X	
J-9	SWIPER Promotion						X				O				
J-10	"Extra Punch" Promotion										O			X	
K-1	First Night Festivities Service			X					X			O	X		
K-2	Holiday Lights Tour			X					X			O	X	X	
K-3	Christmas "Stuff-a-Bus"			X	X							O			
L-1	Special Events Service			X					X				O		
L-2	Off-Peak Promotion			X								X	O		
L-3	Earth Day Celebration		X								X	X	O	X	
L-4	State Fair Traditions		X	X							X		O		

O - Primary Category
X - Secondary Categories

Code	Title of Project/Strategy	A - Accessibility-Related Projects	B - Community Events	C - Cooperative Promotions	D - Image Promotions	E - Internal Promotions	F - Introduction of New Service	G - Media Relations	H - Problem-Solving Projects	I - Promoting Transit	J - Rider Inducements	K - Seasonal Promotions	L - Special Events	M - Target Group Promotions	N - Try Transit Week
L-5	Jackson County Fair Park 'N' Ride		X	X						X			O		
L-6	Mid-State Fair Express Bus Service		X	X									O	O	
M-1	"Behind the Wheel" Campaign				X				X					O	
M-2	Transit Pass for Employees										X			O	
M-3	"Riding Isn't" Campaign									X				O	
M-4	Transit Brochure Distribution									X				O	
M-5	"Summerdime Days"		X	X	X					X				O	
M-6	"The RRTA Senior Game"									X	X			O	
M-7	"Garage Sale Cruise"									X				O	
M-8	Flyers Distributed on the Virginia Tech Campus	X												O	
M-9	Bus Service Guide in Hotels near an Airport													O	
M-10	"The Transit Connection - Connecting the Worker to the Workplace"			X					X	X				O	
M-11	"Class Pass"				X					X	X			O	
M-12	"Smart Tripper Book" and Tour			X				X		X				O	
M-13	"Get On Board"			X	X					X				O	
M-14	School Poster Contest									X				O	X

O - Primary Category
X - Secondary Categories

Code	Title of Project/Strategy	A - Accessibility-Related Projects	B - Community Events	C - Cooperative Promotions	D - Image Promotions	E - Internal Promotions	F - Introduction of New Service	G - Media Relations	H - Problem-Solving Projects	I - Promoting Transit	J - Rider Inducements	K - Seasonal Promotions	L - Special Events	M - Target Group Promotions	N - Try Transit Week
M-15	"Books On Board"			X										O	
M-16	"The Pass"			X						X	X			O	
M-17	Summer Youth Bus Pass									X	X			O	
M-18	"Easy Rider" Student Pass Program			X										O	
M-19	"Library Pass-Time"									X	X			O	
M-20	"Ride 'n' Read" Program			X						X	X			O	
M-21	Metro Benefits Direct Mail Campaign									X				O	
M-22	Direct Mail Piece									X				O	
M-23	Non-Transit Users Survey				X					X	X			O	
M-24	Route-Specific Information Campaign					X				X	X			O	
M-25	Colma New Move-in Mailer					X			X		X			O	
N-1	"Try Transit Day" Promotion		X							X	X				O
N-2	CarFree Days of Spring		X								X				O
N-3	Elementary School Drawing Contest				X									X	O
N-4	Try Transit Week '93		X	X	X					X	X	X		X	O
N-5	Try METRO Week 1997			X	X					X	X				O

O - Primary Category
X - Secondary Categories

Code	Title of Project/Strategy	A - Accessibility-Related Projects	B - Community Events	C - Cooperative Promotions	D - Image Promotions	E - Internal Promotions	F - Introduction of New Service	G - Media Relations	H - Problem-Solving Projects	I - Promoting Transit	J - Rider Inducements	K - Seasonal Promotions	L - Special Events	M - Target Group Promotions	N - Try Transit Week
N-6	Free Ride Coupon and Letter to the Editor										X				O

ANNOTATED BIBLIOGRAPHY

Annotated Bibliography of Marketing Resources

[Note: An attempt has been made to include only those resources that are available as of the date of the publication of this handbook. Details for ordering books or reports from the Technology Sharing Program (TSP), National Technical Information Service (NTIS), and Transportation Research Board (TRB) Bookstore can be found at the end of the Bibliography. For a more complete bibliography on the subject, including publications out of print, please refer to the Final Report for TCRP Project B-13.]

The following annotated bibliography lists resources that contain information that may be useful to the marketing practitioner. Each title is followed by five key words that attempt to capture the essence of the content. The resource is further evaluated for two items, useful information presented and best practices included. The two are rated using a five-star system (★★★★★) with five being the best.

Albrecht, Karl and Zemke, Ron. *Service America: Doing Business in the New Economy*. Homewood, IL: Dow Jones-Irwin. 1985. 203 pp.

Quality Service
Information
Productivity
Involvement
Measurement

Useful Info   ★★★★★
Best Practices   ★★★★★   (non-transit)

Doing business in a service economy calls for a new focus. This focus must be customer driven. From the management of the company to the provision of service, the entire team must be in step with expectations, research, measurement, motivation, problem solving, and quality improvement. A classic in the customer-driven marketing revolution. [Available in major bookstores]

Anderson, Kristen and Zemke, Ron and Performance Research Associates, Inc. *Delivering Knock Your Socks Off Service*. New York: AMACOM. 1991. 136 pp.

Customers
Needs
Listening
Criteria
Perception

Useful Info   ★★★★★
Best Practices   ★★   (non-transit)

An easy to read, delightful discussion of the basics to delivering service that your customers want. From defining customers as everyone who interfaces with the company to doing your job well, thoughtfully, skillfully, and to the customer's delight, this book shares the secrets of thousands of quality-oriented customer service professionals. [Available in major bookstores]

Anderson, Kristen and Zemke, Ron and Performance Research Associates, Inc.
Knock Your Socks Off Answers: Solving Customer Nightmares & Soothing Nightmare Customers. New York: AMACOM. 1995. 143 pp.

Customers
Expectations
What to Say
Attitude
Problem Definition

Useful Info   ★★★★
Best Practices   ★★★★ (non-transit)

This book, the fourth in the series, provides the guidelines on how to show poise in service delivery; how to conduct a negotiation in an atmosphere of partnership; how to give the customer a sense of security, a sense of control over the situation; how to use a good working theory of what you want to accomplish with the customer, once you know the wants, needs, attitudes, hopes, and fears of the individual. [Available in major bookstores]

Anderson, Kristen and Zemke, Ron and Performance Research Associates, Inc.
Tales of Knock Your Socks Off Service: Inspiring Stories of Outstanding Customer Service. New York: AMACOM. 1998. 193 pp.

Above and Beyond
Caring
Sincerity
Service Sensitivity
Commitment

Useful Info   ★★★★★
Best Practices   ★★★★★ (non-transit)

The sixth book in the popular series on customer service, eschews the straightforward advice and direct suggestions dispensed by its predecessors and introduces 175 real-life examples meant to propel readers toward similar conduct. Anecdotes are divided into themed sections, such as "Above and Beyond" (celebrating actions that "surpass the call of duty") and "Great Saves" (spotlighting "recoveries that renew customer faith"), and offer a half-dozen profiles of exemplary corporations, including Lands' End and Norwest Bank. The book shares shining examples of the "Daily Delights," the "Great Saves," and the "Random Acts of Service Kindness" that make customers remember — and that bring them back to an organization again and again. With its humor, pragmatic observations, and dozens of stories from "the service zone," anyone at any service level will get a kick out of this book. (And learn some lessons along the way!) [Available in major bookstores]

Bell, Chip and Zemke, Ron and Performance Research Associates, Inc.
Managing Knock Your Socks Off Service. New York: AMACOM. 1992. 310 pp.

    Process Control
    Culture
    Strategies
    Employees
    Empowerment

    Useful Info   ★★★★★
    Best Practices   ★★   (non-transit)

The second book in a series, it provides the tools needed to manage the process. It describes the strategies of developing a culture in your organization. Here the emphasis is on the employees who serve as the front line representatives of your company. [Available in major bookstores]

Cambridge Systematics, Inc., et al. *Transit Marketing: A Program of Research, Demonstration, and Communication.* UMTA-MA-06-0049-85-8. 1985. 72 pp.

    Research
    Demonstration
    Communication
    Techniques
    Evaluation

    Useful Info   ★★★
    Best Practices   ★★

This report recommends a five-year program of research, demonstration, and communication to improve the effectiveness of marketing practice in the U.S. transit industry. The program is oriented toward the development of improved market research tools, strategic planning, marketing techniques, and evaluation methods. The program includes two phases - (1) research and development, (2) deployment of prototypical marketing programs based on the results of the research program. Training and information dissemination help translate the research efforts. [Available through NTIS]

Carter, Goble, Roberts, Inc. *South Carolina Transportation Marketing Manual.* Washington, D.C.: U.S. DOT-Technology Sharing. DOT-1-84-18. September 1980. 194 pp.

> Positioning
> Action Plan
> Evaluation
> Implementation
> Image

> Useful Info   ★★★★
> Best Practices   ★★

> This manual was prepared to aid the transit systems of South Carolina with some of the best transit marketing knowledge, examples, and procedures found in the U.S. The manual provides a systematic approach to the marketing management function with the primary focus being on better service to the customers. It is designed to cover planning, development, execution, and refinement of a marketing program. It also contains several examples of transit promotional material. [Available through NTIS]

Centre for Transit Improvement. *A Handbook for Transit Managers — Promotions, Publicity, and All That Pizazz.* Ontario Urban Transit Association. 1988. 100 pp.

> Promotions
> Best Practices
> How To
> Community Service
> Ridership

> Useful Info   ★★★★★
> Best Practices   ★★★★★

> An easy-to-read how-to guide designed to transfer transit promotions that work. From objective to strategy to materials to results, each project is briefly described. Several categories are used in promoting transit: rider inducements, image/fundraising, co-op promotion, group promotions, solving problems, seasonal promotions, involving employees. A recipe book for cooking up transit promotions. [Available through the Ontario Urban Transit Association's Centre for Transit Improvement, 55 York Street, Suite 901, Toronto, Ontario, Canada M5J 1R7, 416/365-9800]

Communiqué, L.L.P. *The Marketing Cookbook: "Recipes for Success"*. Denver, CO: Colorado Association of Transit Agencies. 50 pp.

    Basics of Marketing
    How-To
    Program Planning
    Creative Samples
    Evaluation

    Useful Info   ★★★★★
    Best Practices ★★

This is a straightforward, simple overview of the basics of marketing applied in the public transit setting. It is an excellent resource for "learning the ropes" of marketing, or reminding seasoned professionals of the basics. Handy forms for developing a marketing program, or executing a research program (survey instruments) are included. Useful samples are also included for media releases, public service announcements, advertising campaigns and radio spots. [Available through the Colorado Association of Transit Agencies, 225 East 16th Avenue, Suite 1070, Denver, CO 80203, 303/839-5197]

Connelan, Thomas K. and Zemke, Ron and Performance Research Associates, Inc. *Sustaining Knock Your Socks Off Service*. New York: AMACOM. 1993. 176 pp.

    Program Development
    Management
    Employees
    Training
    Feedback

    Useful Info   ★★★★★
    Best Practices ★★   (non-transit)

This book, the third in a series, provides a journey on not only how to set up quality customer service programs but also how to maintain them. Just as research shows that good companies continually try to get better, while mediocre ones seldom try anything to keep from getting worse, this book provides a road map on how to develop that service edge. [Available in most bookstores]

Everett, Peter B. and Shoemaker, Robert N. and Doscher, William E. *Public Transportation Marketing Evaluation Manual: Techniques for Data Collection.* Washington, D.C.: U.S. DOT Technology Sharing. DOT-T-88-06. October 1987. 82 pp.

Data Collection
Information Aids
Pricing
Advertising
Measurement

Useful Info ★★★
Best Practices ★

The purpose of this manual is to suggest ways to reconstitute and make more vigorous the marketing function in transit. It suggests three ways of accomplishing this goal: (1) shift the emphasis of public transit to community transit with a family of service alternatives, (2) reposition within the public transportation organizational structure, and (3) utilize effective evaluation methods which is the central theme of this manual. It suggests moving away from some of the efficiency measures and incorporating effectiveness measures in the evaluation process. [Available through TSP]

Gigante, Lisa and Koo, Emily. *Marketing Public Transit: An Evaluation.* Washington, D.C.: U.S. DOT-UMTA. 1985.

Attitudes
Research
Surveys
Awareness
Innovation

Useful Info ★★★★★
Best Practices ★★★★

In 1983, Michigan DOT received a grant of $335,000 to be used in the development of innovative marketing techniques which it delivered to 11 systems. This report describes the projects and provides recommendations for the implementation of similar techniques by other transit systems. In particular, a research project was able to determine the effectiveness of promotional activities on the public's awareness, attitude, and usage of the local bus systems. [Available through the NTIS]

Grey Advertising, Inc. and Caruso, Joe, ed. *Transit Marketing Management.*
UMTA-URT-30-88-1. Mar 1988. 60 pp.

Transit Marketing
Survey
Research Guidelines
Techniques
Problem solving

Useful Info ★★★
Best Practices ★

This report offers a broad overview of consumer research as it applies to specific problems of marketers in transit. This research should be "customer-oriented," delineate market segments for target marketing, solicit and interpret a wide range of attitudinal data from current and potential customers, and investigate the relationship between stated intentions and actual behavior of these individuals. It also urges a standardization and coordination of data gathering instruments and procedures, which would enable the collection of more accurate data. [Available through NTIS]

Ilium Associates, Inc. *A Handbook for Effective Advertising and Marketing of Community Transit.* Bellevue, WA: CTAA. 1989. 50 pp.

Marketing Planning
Promotions
Public/Community Relations
Rural Transit
Creative Materials

Useful Info ★★★★
Best Practices ★

This guide was developed under contract with the Community Transportation Association of America and aimed at rural and community transit systems. It is a mix of worksheets to be used in planning the marketing program and hints and tips about advertising, promotions, public and community relations, and generating revenue. [Available from CTAA - 1341 G. Street, N.W., Suite 600, Washington, D.C. 20005, 800-527-8279]

Knapp, Sue, Ecosometrics Incorporated, and Wisconsin DOT. *Marketing Manual for Shared-Ride Taxi Systems in Wisconsin*. Washington, D.C.: U.S. DOT Technology Transfer (Reprint Series). DOT-1-87-25. February 1987. 78 pp.

Ridership
Community Support
Planning
Market Description
Implementation

Useful Info   ★★★★
Best Practices   ★★★★

Even though the purpose of this manual is to assist local shared-ride taxi systems in preparing and implementing marketing programs, it is also an excellent reference for small transit systems to use for the same purpose. The report is easy to read and presents a comprehensive flow of "how-to" activities from an overview of marketing to the community to a list of potential marketing activities organized by function. The suggested marketing strategy sets up guidelines for concentrating resources and efforts on likely users. [Available through NTIS]

Kotler, Philip and Andreasen, Alan. *Strategic Marketing for Non-Profit Organizations*. 5th ed. Englewood Cliffs, NJ: Prentice-Hall, Inc. 1995. 528 pp.

Exchanges
Research
Benefits
Customer
Social Marketing

Useful Info   ★★★★★
Best Practices   ★★★★★ (non-transit)

"The question is not whether one will use marketing but whether one will use it better than one's competitor or than one has in the past.... This means knowing the most sophisticated and advanced tools and concepts the field has to offer and being able to use them on a day-to-day basis." This quote from the preface of this textbook provides a glimpse of its comprehensive nature on the subject. The author's treatment of services marketing has many direct applications for transit marketing. [Available through major bookstores]

MORPACE International, Inc., and Cambridge Systematics, Inc. *Customer-Defined Transit Service Quality. (Final Draft Report).* Washington, D.C.: TRB, National Research Council TCRP B-11. January 1998. 75 pp.

Customer Satisfaction
Customer-Defined Quality
Drivers of Satisfaction
Attribute Impacts
Things Gone Wrong

Useful Info   ★★★★★
Best Practices   ★

This reports builds on the Customer Satisfaction Index (CSI) for the Mass Transit Industry (1995). Addressing the challenge that transit agencies need reliable and official methods for developing customer satisfaction benchmarks and tracking indices, this report suggests a simpler approach. It suggests a method for measuring the impact of individual service quality attributes on overall satisfaction. They call the method "things gone wrong" or a "problems encountered" approach. This approach relies on the concept that overall satisfaction is negatively impacted when a customer experiences a problem with a service attribute that is a driver of satisfaction. It presents the (1) rationale/strength of this approach, (2) a guide for conducting "things gone wrong" analysis and tracking results, (3) field test data to illustrate the usefulness of the approach, and (4) a process for monitoring customer satisfaction. [Available from the TRB Bookstore]

Multisystems, Inc., et al. *Effective Methods of Marketing Transit Services to Business (Interim Report).* Washington, D.C.: TRB, National Research Council TCRP B-8. December 1996. 74 pp.

Business to Business
Goals
Needs
Target Segmentation
Research

Useful Info   ★★★
Best Practices   ★★★

This report reviews the experiences of transit and non-transit organizations in their efforts to implement business-to-business marketing. Following the grand scheme of testing, refining, implementing, and evaluating a marketing program, the report describes strategies to meet needs and satisfy customers. It offers many examples of where specific target markets need specific service alternatives. It draws examples from public transit, the banking and travel industries, and non-profit organizations. The seven major elements are used as a guide in the review of approximately 50 transit properties with business-to-business programs. [Available through the TRB Bookstore]

Northwest Research Group, Inc. *A Handbook: Integrating Market and Customer Research Into Transit Management.* TRB, National Research Council TCRP Report 37, 1998, 207 pp.

Research
Management
Customer
Strategies
Measurement

Useful Info ★★★★
Best Practices ★★★

A "must" handbook for the library of the person interested in research and customer satisfaction. It is delightfully written, well-organized, comprehensive, and easy to use. Nice features: a listing of available tools in the toolbox at the beginning of each chapter; "road maps" used frequently to emphasize key elements; complete and helpful information in each of the chapters. Twelve case studies help to visualize the application of many of the marketing activities. The glossary reinforces the excellent material found in the text. [Available through the TRB Bookstore]

Northwest Research Group, Inc. *A Handbook: Using Market Segmentation to Increase Transit Ridership.* TRB, National Research Council TCRP Report 36, 1998, 194 pp.

Research
Methods
Procedures
Positioning
Mix Application

Useful Info ★★★★
Best Practices ★★★

This handbook presents a refreshing perspective on market segmentation without a heavy reference to equations and complex formulations. It reviews the basic terms and methods that are used in performing market segmentation analysis. The report presents the science and how it could be used in an organization's strategy. It can be used as a guide for that manager who needs to better understand the results. It can also be used as a checklist for the market researcher assigned with the task of implementation. [Available through TRB Bookstore]

Peter Muller-Munk Associates (Division of Wilbur Smith and Associates).
Transit Marketing in Pennsylvania: A Handbook of Effective Transit Marketing Aids.
Washington, D.C.: U.S. DOT Technology Sharing Program. DOT-1-81-36. 1984. 60
pp.

    User Information
    Promotion/Advertising
    Fares
    Public Relations
    Service Planning

    Useful Info   ★★★
    Best Practices ★★★★

This handbook was developed to help smaller transit systems. The purpose is
to attempt to satisfy the need for a precise, comprehensive information
exchange on the availability and usage of transit marketing items in
Pennsylvania. In order to extract the maximum benefits from past marketing
efforts, the aids and promotions are evaluated for perceived benefits and effectiveness. [Available through NTIS]

Schmied, Beatrice, et al. *Promotions, Publicity, and All That Pizazz (Round Two).*
Ontario Urban Transit Association. 1994. 111 pp.

    Promotions
    Best Practices
    How-to
    Planning
    Media

    Useful Info   ★★★★★
    Best Practices ★★★★★

"Round Two" is an updated version of the first publication. Welcome additions are chapters on environment, system information, accessibility, and revenue raiser projects. Also included are suggestions for planning an event and a
brief list of tips for dealing with the media. The illustrations and photographs
are also pleasant additions that may help to get the message across. [Available
from the Ontario Urban Transit Association's Centre for Transit Improvement,
55 York Street, Suite 901, Toronto, Ontario, Canada M5J 1R7, 416/365-9800]

Stern, Gary J. *Marketing Workbook for Non-Profit Organizations*. St. Paul: Amherst H. Wilder Foundation. 1990. 132 pp.

Value
Goals
Audit
Plan
Promotion

Useful Info   ★★★★★
Best Practices   ★★★   (non-transit)

This is a wonderful guide to basic marketing that any organization can put to use. The reader is apt to learn about marketing as a creative enterprise undertaken with a thinking and caring spirit. Transit marketers are likely to gain insights that lead to promoting value, accomplishing mission, and developing resources that can help address a range of concerns. The fill-in-the-blank marketing plan worksheets are well worth the price of the workbook. [Available from the Foundation - 919 Lafond Avenue, St. Paul, MN 55104, 800/274-6024]

Tri-County Metropolitan Transportation District of Oregon. *Customer Satisfaction Index for the Mass Transit Industry*. Washington, D.C.: National Research Council, TRB IDEA Program. 1995. 40 pp.

Customer Satisfaction Index
Key Satisfaction Drivers
Data Collection
Comparison Methods
Research

Useful Info   ★★★★★
Best Practices   ★★★

This pilot customer satisfaction index (CSI) research project is the first systematic unbiased, statistically sophisticated measure of customer satisfaction to be conducted across transit districts. For the first time, transit agencies have the ability to analyze their own performance, compare themselves directly to a total sample average, and compare and learn from other districts. Five transit properties participated in this project. The results from the project successfully illustrated the feasibility of using a common measurement of customer satisfaction to assist decision makers in transit to achieve a better fit between the features of transit services and the needs of the customers. The features that were uncovered included: (1) courtesy of drivers; (2) availability and comfort of seats; (3) frequency of service; (4) safety from crime; and (5) cleanliness of the vehicles and bus stops. It is important to note that the safety/security issues are intricately tied to cleanliness. The study report provides suggestions for doing the following: (1) learning from transit properties identified as generating excellent customer satisfaction results; (2) creating and promoting customer satisfaction programs; (3) comparing data collection techniques; (4) directing budget expenditures to transit features that make a difference; and

(5) reinforcing an agency's ability and commitment to match services to customer needs. [Available through the TRB Bookstore]

Utah Transit Authority. *Celebrating Excellence in Public Transit*. Washington, D.C.: U.S. DOT Office of Technical Assistance and Safety, Federal Transit Administration. DOT-T-93-19. 1993. 229 pp.

Transit Excellence
General Operations
Best Practices
Customer Service
Empowerment

Useful Info ★★★
Best Practices ★★★★

This report documents an FTA-funded effort by the Utah Transit Authority to study excellence in the transit industry. It showcases excellence in the industry, and calls for actions to pursue innovation and excellence. A complimentary video tape showcased 12 of the exemplary projects. While overall transit industry excellence is covered, there are several sections calling attention to successful marketing projects. Topics include: community involvement and leadership; customer service; empowerment/participation/teams; marketing; quality management; reward and recognition; safety/health/fitness; service design; technology; and values and cultures. [Available through TSP]

Walb, Carol A. and Loudon, William R. and Cambridge Systematics, Inc.
Transit Marketing: A Review of the State-of-the-Art and a Handbook of Current Practice.
UMTA-MA-06-0049-85-7. 144 pp. 1985.

Transit Marketing
Market Research
Marketing Techniques
Best Practices
Ridership

Useful Info ★★★★
Best Practices ★★★★

This report attempts to relate the use of marketing to improved transit ridership and productivity. It is an overview of current practices (1983-1984) found in 25 transit agencies across the nation. It includes a discussion of marketing's role and presents a framework for a comprehensive approach. The appendix presents additional detail and examples of marketing activities at individual transit agencies. [Available through NTIS]

West Virginia Public Transportation Division. *Transit Marketing Manual: Get on the Bus and Ride*. Washington, D.C.: U.S. DOT Technology Sharing Program (Reprint Series). DOT-I-85-23. 1984. 73 pp.

Newspapers
Radio
Public Relations
Public Service Announcements
Education

Useful Info ★★
Best Practices ★★

This marketing handbook is the official collection of ad materials from the West Virginia Public Transportation Division. Its purpose is to aid in making the most of the advertising dollar, to help expand public influence, and to increase ridership in local areas. Some of the illustrations, scripts, press releases, and advice may be helpful. [Available through the TSP]

Wisconsin Urban Transit Association & WISDOT. *Wisconsin Transit System Marketing Manual*. Washington, D.C.: U.S. DOT Technology Sharing Program (Reprint Series). DOT-I-87-26. March 1987. 212 pp.

Mix
Planning
Implementation
Best Practices
Promotion

Useful Info ★★★★
Best Practices ★★★★

A "how-to" guide designed to take the reader through the marketing planning process step by step. It was designed to assist in developing specific advertising and promotion plans to increase ridership. The "Ideas Grid by Strategy" provides 88 ideas related to 16 strategies from motivating employees to enhancing your image. The advertising slicks provide a camera ready art form that can be helpful to any transit system. [Available through NTIS]

Zemke, Ron and Anderson, Kristin and Performance Research Associates, Inc.
Coaching Knock Your Socks Off Service. New York: AMACOM. 1997. 148 pp.

Preparation
Feedback
Performance
Positioning
Closure

Useful Info   ★★★★★
Best Practice   ★★   (non-transit)

This book addresses the basic philosophy and skills of effective coaching. It further offers guidelines regarding coaching the new employee for high performance, coaching without preparation, coaching the unsure employee, coaching for difficult duty and special situations, coaching trouble on the team, and finally peer coaching. This book contains great common sense approaches to developing excellent employees. [Available in major bookstores]

Zemke, Ron and Schaaf, Dick. *The Service Edge: 101 Companies That Profit From Customer Care.* New York: NAL Penguin, Inc. 1989. 584 pp.

Survival
Expectations
Customers
Monitoring
Select/Train

Useful Info   ★★★★
Best Practices   ★★★★ (non-transit)

This book provides a menu of exemplars to examine as we consider the role of customer-driven service. The richly detailed descriptions of service practices in these diverse companies help the reader settle in on the common denominators of good quality service. [Available in major bookstores in paperback]

TSP/NTIS/TRB Contact Information

Many of the documents mentioned in this "how-to" book are available through the U.S. Department of Transportation's Technology Sharing Program (TSP). These documents were developed with direct or indirect support of federal funds. Single copies of in-stock TSP reports are available at no cost through the main on-line catalog accessed through the Internet at http://www.tsp.dot.gov/. TSP can be contacted also at Order Desk - DRA-4, U.S. DOT, 400 Seventh Street, SW, Washington, DC 20590. 202/366-4999 fax - 202/366-3272.

All TSP reports are archived through the National Technical Information Service (NTIS) in Springfield, VA. When reports are no longer available through DOT sources, reports may be purchased through NTIS. Archived TSP reports may be browsed through the TSP web page, and a direct link to NTIS is provided. NTIS may be contacted directly at: NTIS Sales Desk, Monday through Friday, 8:30 a.m. to 8:00 p.m. Eastern Time, at 800/533-6847 or 703/605-6000, fax 703/321-8547. Additional NTIS ordering information is available at the NTIS website - http://www.fedworld.gov/ntis/ordering.htm.

Other documents, particularly those sponsored by the Transit Cooperative Research Program, are available through the Transportation Research Board (TRB) Bookstore. Purchases may be made by mail, fax, phone or online. For additional information, see the Bookstore's website at http://www2.nas.edu/trbbooks/. TRB may also be contacted at: TRB Bookstore, P.O. Box 289, Washington, DC 20055, 202/334-3214, fax - 202/334-2519.